THE CHANGING NATURE OF ECO/FEMINISM

THE CHANGING NATURE OF ECO/FEMINISM

TELLING STORIES FROM CLAYOQUOT SOUND

Niamh Moore

UBCPress · Vancouver · Toronto

22 21 20 19 18 17 16 15 5 4 3 2 1

Printed in Canada on FSC-certified ancient-forest-free paper (100% post-consumer recycled) that is processed chlorine- and acid-free.

Library and Archives Canada Cataloguing in Publication

Moore, Niamh, author

The changing nature of eco/feminism : telling stories from Clayoquot Sound / Niamh Moore.

Includes bibliographical references and index.
Issued in print and electronic formats.
ISBN 978-0-7748-2627-3 (bound). – ISBN 978-0-7748-2628-0 (pbk.). –
ISBN 978-0-7748-2629-7 (pdf). – ISBN 978-0-7748-2630-3 (epub)

1. Ecofeminism – British Columbia – Clayoquot Sound. 2. Feminists – British Columbia – Clayoquot Sound – Interviews. 3. Women political activists – British Columbia – Clayoquot Sound – Interviews. 4. Clayoquot Sound (B.C.) – Environmental conditions. I. Title.

HQ1194.M65 2015 305.4209711'2 C2013-903397-1
C2013-903398-X

Canada

UBC Press gratefully acknowledges the financial support for our publishing program of the Government of Canada (through the Canada Book Fund), the Canada Council for the Arts, and the British Columbia Arts Council.

This book has been published with the help of a grant from the Canadian Federation for the Humanities and Social Sciences, through the Awards to Scholarly Publications Program, using funds provided by the Social Sciences and Humanities Research Council of Canada.

UBC Press
The University of British Columbia
2029 West Mall
Vancouver, BC V6T 1Z2
www.ubcpress.ca

Forest

THE WAY WE STAND

The way we stand, you can see we have grown up this way together, out of the same soil, with the same rains, leaning in the same way toward the sun. See how we lean together in the same direction. How the dead limbs of some of us rest in the branches of another. How these branches have grown around the limbs. How the two are inseparable. And if you look you can see the different ways we have taken this place into us. Magnolia, loblolly bay, sweet gum, Southern bayberry, Pacific bayberry; wherever we grow there are many of us; Monterey pine, sugar pine, white-bark pine, four-leaf pine, single-leaf pine, bristle-cone pine, foxtail pine, Torrey pine, Western red pine, Jeffry pine, bishop pine. And we are various, and amazing in our variety, and our differences multiply, so that edge after edge of the endlessness of possibility is exposed. You know we have grown this way for years. And to no purpose you can understand. Yet what you fail to know we know, and the knowing is in us, how we have grown this way, why these years are not one of them heedless, why we are shaped the way we are, not all straight to your purpose, but to ours. And that we each have a purpose, how each cell, how light and soil are in us, how we are in the soil, how we are in the air, how we are both infinitesimal and great and how we are infinitely without any purpose you can see, in the way we stand, each alone, yet none of us separable, none of us beautiful when separate but all exquisite as we stand. Each moment heeded in this cycle, no detail unlovely.

– SUSAN GRIFFIN, ***WOMAN AND NATURE*** (1978)

the silence of trees
the silence of women

if they could speak an unconditioned language
what would they say

– DAPHNE MARLATT, *ANA HISTORIC* (1990)

My physical mother is gone. My spiritual mother remains. I am a woman rewriting my genealogy.

– TERRY TEMPEST WILLIAMS, *REFUGE: AN UNNATURAL HISTORY OF FAMILY AND PLACE* (1991)

Contents

Preface

"She Goes On and On and On"

> You can't kill the Spirit
> She is like a mountain
> Old and strong
> She goes on and on and on
> She is like a mountain ... (repeat ad nauseam)
>
> – NAOMI LITTLEBEAR MORENA, "YOU CAN'T KILL THE SPIRIT"

I was struck by the phrase "repeat ad nauseam," when I went looking on-line for the words of the song "You Can't Kill the Spirit." I hadn't really forgotten the lyrics, having sung them so many times at the blockades that hoped to prevent MacMillan Bloedel from clear-cutting Clayoquot Sound during the few weeks I was there. Perhaps it was more a case of a researcher needing something more tangible than memory, a written trace of the words – on-line and virtual and so, real and right. Mainly, of course, avoiding finishing, perhaps even avoiding starting, a (proper) sentence. Somewhere I do have an original song sheet from the camp, which was already a copy, a photocopied page of recycled paper with the lyrics most frequently sung on the blockades. Writing out the songs and making copies for everyone was a suggestion from Sue, who turned out to be an anthropologist. Her neat writing is somewhere in my research archive now, but it's so much easier to go on-line than to open that Pandora's box; who knows what else is in there that I have forgotten about – or remember only too intensely. I was relieved when I eventually came across Geoff Bowker's (2005, 15) acknowledgment of all the notes that he has taken, knowing they will never be looked at again. Memory and headnotes (see Ottenberg 1990), not so much field-notes, most often guide this writing.

On my computer screen, a myriad of connections unfolded. The song, described as a spiritual and a round, was included as part of a Greenham Common Women's Peace Camp Songbook. And though I wasn't at Greenham, I knew the words before I went to Clayoquot and am sure that I sung them later at Newbury too, with Margaretta, who had been at Greenham, and Seneca as well. At Clayoquot, I sung them with Sue and her partner, Carol, who had been to Greenham, and with hundreds of others; some, like Chris from the Friends of Clayoquot Sound, had also been to Greenham.

The website that came up first – of an independent voluntary organization called the Danish Peace Academy – named Naomi Littlebear Morena, Chicana feminist musician, as author and composer of the song. On the website, which offers a wonderful archive of peace activism, there was also an extract from *Spare Rib,* a British feminist magazine, featuring an interview with Morena by someone from the Manchester Greenham Support Group regarding her planned trip to sing the song at Greenham once she had found out that women there were singing her song. Later, I found out that the words "You can't kill the Spirit" are written into the spiral sculpture of steel and stone at the commemorative site that marks nineteen years of continuous peaceful occupation of the Greenham women's camp.[1]

"You Can't Kill the Spirit" was sung ad nauseam during the dim mornings at the Clayoquot blockade, as we gathered at the road into the logging area. Some of us were preparing to be arrested; others, like me, were still plucking up the courage; and some brave (mad?) people, such as Síle Simpson, were refusing to stay away from the road, even when she had been arrested and was not supposed to return there. Síle was arrested three times that summer, so clearly the injunction didn't stop her; in the end, she was sentenced to six months of electronic monitoring, but that didn't seem to stop her either.

Kami Kanetsuka, whom I met on the bus from Nanaimo, had come to the camp because, like a contemporary Mona Lisa, Síle had gazed at her from a newspaper cutting on her fridge door gallery, eyes following her around the kitchen as she went about her daily life, all day, every day, all summer long (ad nauseam?). Eventually, she felt she had to put on her backpack and set off for the camp.

For some, the ad nauseam was almost literal. They had been at the camp all summer, and who knows how many times they'd sung it. They were clearly sick of going on and on and on, and they groaned aloud when someone suggested singing it again. Others, like me, were still a little new to it all and hadn't been there long enough to become sick of the repetition, though perhaps we were a little uncomfortable with the singing – I didn't, and still don't, *sing*, except a bit of chanting at yoga now. I did feel a bit sick with the weirdness of how I had ended up on a road at 4:00 a.m., in the early morning light, in the "company of strangers" (see Meigs 1991; C. Scott 1991). But I was also mildly comforted by the collective repetition. It was barely singing, after all, so short and so repetitive that even I could manage it. And it helped to calm me, as I stood on the road, with the decision hanging over my head to "choose" to be arrested, sentenced, stopped in my tracks.

And in the end, being arrested was no decision, no effort at all, just a giving up of the effort, of the struggle; I was arrested by the trees, and the women, and the shared practice of eco/feminist politics long before the Royal Canadian Mounted Police (RCMP) got to me.

And perhaps, for the Friends of Clayoquot Sound (FOCS), the camp itself was an ad nauseam – threatening to go and on and on until, paradoxically, the FOCS felt that it must be brought to an end if the work of campaigning were to be sustained.

I first heard of Clayoquot Sound and the campaign against clear-cut logging through the Women's Environmental Network (WEN) in London in 1992. It came up during a fleeting conversation with a WEN activist, I think her name was Helen Moss, the day before she left for Vancouver to start work with Greenpeace. At that moment, as she was trying to empty her desk, she probably did not need an enthusiastic student showing up unannounced, but when I asked whether anything "eco/feminist" was going on in Canada, she gave me a few minutes of her precious time and dug out an *Observer Magazine* article about Clayoquot (Lean 1991). I still have it, but dare I confess that I remember being disappointed at the time? I mean, trees were nice, but hardly – what – interesting? I'm not sure now what I was expecting, or perhaps hoping for. I think maybe I had dirtier things in mind: nuclear and chemical accidents (Sellafield, Three Mile Island, Bhopal), pollution, toxic waste, cancers, death, destruction – the kinds of things that wilderness and trees didn't quite conjure up for me. At least, not then. Perhaps these were not quite the right desires for an aspiring eco/feminist, but somehow they felt familiar, and wilderness did not. Now, after hearing horror stories about MacMillan Bloedel and seeing clear-cuts for myself, after reading histories of forestry and of Vancouver Island, and learning more of genocide and ecocide, after hearing the stories of those who campaigned against clear-cutting and having spent time in Clayoquot, including a salutary few minutes alone in a jail cell, I begin to appreciate just how dirty clear-cut logging can be. And I appreciate trees more, though I'm sure too that many readers might hope for more appreciation than will appear in these dead dry leaves. But this book is not so much about wilderness or rainforest as abstract, naturalized, or romanticized entities; rather, it is about Clayoquot, about "situated knowledges," "the politics of location," "the friction of global connections," "shadow places and the politics of dwelling" (Haraway 1988; Rich 1987; Tsing 2005; Plumwood 2008).

My enquiry to WEN about eco/feminism in Canada was not only about a desire to be with other eco/feminists who were saving the planet. Rather, more pragmatically, the Centre of Canadian Studies at Queen's University Belfast, where I was doing an undergraduate degree in women's studies and politics, was offering an annual travel award of a trip to Canada to do a small research project, and I thought I might apply. At the end of the application form, I scribbled something about seeing whether the Clayoquot campaign included any explicitly eco/feminist dimensions, and some months afterward, I found myself wending my way to Canada, to a campaign that had meantime transformed into blockades, supported by a peace camp. My experiences there profoundly influenced my understanding of feminism and continue to preoccupy me many years later.[2] It was much later too before I realized it was no accident that WEN led me to Canada, or that Clayoquot led back to WEN, because WEN was the only major environmental organization in Europe with a campaign around temperate rainforests during the early 1990s; most others were focused on tropical rainforests such as in the Amazon basin.

I came back from Vancouver Island in the late summer of 1993 to the final year of my undergraduate degree in women's studies and politics, to write an account of eco/feminist activism in Clayoquot for the Centre of Canadian Studies as a requirement of the travel award and for my undergraduate dissertation. I came back to lively courses on women and politics – ten weeks, from Mary Wollstonecraft (1792) and Mary Shelley (1818) to a final session on Camilla Paglia (1990), Naomi Wolf (1991), and Susan Faludi (1992). After an intense, vibrant, and lively experience of eco/feminist politics in action, I came back to books and media stories of the end of feminism, of backlash, and of the rise of postfeminism. It was disorienting – how was I to account for the dissonance between these books and my own experiences? Even at the time, I think I understood that Wollstonecraft and Shelley offered more resources for thinking about eco/feminist activism than the triumvirate of Paglia, Wolf, and Faludi. Increasingly, it seemed that the terms "essentialism" and "universalism" were the only ones available for thinking about eco/feminism. Yet such accounts did not cohere with my own nascent passions about eco/feminism, feminism, and environmentalism, with my experiences of eco/feminism in Clayoquot, or with my readings of the emerging literature on eco/feminism.

Later, I came across Lee Maracle's book, *I Am Woman,* and her words: "There is a story in every line of theory, not in our capacity to theorize. It seems a waste of words to dispassionately delete character from plot line, tension and conclusion. It takes a great deal of work to erase people from theoretical discussion" (1996, 236). This seemed to describe so much of the literature that reduced women's activisms and lives to essentialism, one word for such vast experiences and stories. So in the summer of 1996, I travelled again to British Columbia, this time to do doctoral research. Transforming experience into research is typical of feminist work. As Beverley Skeggs (1995, 12) writes, "As feminists we are involved in continually theorising the situations and locations of ourselves and others, we are already implicated before we begin research." Optimistically (and now I add ambitiously, naively), I wanted to counter reductive readings of eco/feminism, to resist collapsing it into essentialism or universalism. I wanted to open out these stories, which seemed so curtailed and truncated by a prevailing reliance on essentialism and universalism, and that seemed to dismiss many of eco/feminism's passions. I wanted to record and (re)tell stories of women and trees.

> We are eating up the planet, the *New York Times*
> takes a forest, every Sunday, Los Angeles
> draws its water from the Sacramento Valley
> the rivers of British Columbia are ours
> on lease for 99 years.
>
> – DIANE DI PRIMA, *REVOLUTIONARY LETTERS* (2007)

Now, two decades later, I turn to Clayoquot to tell stories about eco/feminism and in the process to undo other stories that are told about eco/feminism, feminism, and Clayoquot itself. As with Anna Tsing's (2005, ix) account of forestry in Indonesia, "it is not a story which can be confined in a village, a province or a nation," but one that is densely entangled with global eco/feminisms, the travels of women and trees. As Tsing writes in *Friction* (ibid., 1-2), her ethnography of global connection, "The specificity of global connections is an ever-present reminder that global claims do not actually make everything everywhere the same ... As soon as we let go of the universal as a self-fulfilling abstract truth, we must become embroiled in specific situations. And thus it is necessary to begin again, and again, in the middle of things." Although eco/feminism has often been dismissed for its supposed universalizing tendencies, for implying a unity between women across the world, and for reifying certain Third World movements, such as the Chipko Movement, Clayoquot seemed a useful site for undoing such simplifications. As the FOCS easily understood, Clayoquot is deeply implicated in the rest of the world. For those who might puzzle over where Clayoquot is, finding a small dot on the map of Vancouver Island does not necessarily offer much of an answer to the where, or what, of Clayoquot. It is more useful to follow Warren Magnusson and Karena Shaw, who recognize that Clayoquot cannot be so easily located or confined. In their thoughtful book *A Political Space: Reading the Global through Clayoquot Sound,* Magnusson and Shaw (2003c, vii, emphasis in original) note that Clayoquot "appears to be at the *periphery* of contemporary power and authority and thus to be *marginal* to the study of politics." However, as they (ibid.) suggest, "This appearance is an effect of particular assumptions, assumptions that need to be challenged. Clayoquot is more usefully interpreted or 'read' as a center of interaction among the movements, powers, and authorities that produce the world in which we live." Neither can Clayoquot be reduced to any narrow understanding of environmental politics, as Magnusson and Shaw (ibid., viii, emphasis in original) reflect: "The more research we have done, the more interesting Clayoquot has become to us as a place where *other* aspects of contemporary politics – be they gender relations or urban-development conflicts or struggles over the nature of community – come into focus."

With this opening, I gesture toward a genealogy of this book. Over twenty years have passed since the summer of 1993. The FOCS continues to campaign. Stories of the end of feminism persist, even in the face of the many ongoing manifestations of feminism. There are now other stories of the recent feminist past, and present, and future. There is intense interest in (re)writing feminist history. This comes in part from accounts of third and fourth waves, and from anxieties about generational conflicts. The emergence of "new material feminisms" offers a further commentary on the recent feminist past, if not always explicitly. At best, eco/feminism is relegated to a footnote in some of these histories, and mostly it does not figure at all. The insistent opposition to essentialism, the adherence to particular notions of anti-foundationalism, the desire to demonstrate that white feminists finally understand the problems of global sisterhood, and the baseless assumption that eco/feminism falls foul of all of these have meant that finding a way to write about eco/feminism has been difficult. Yet in the context of a renewed attention to the politics of feminist histories, at the same time as some insistently name new material feminisms against some imagined history that never addressed the material adequately, this is perhaps a moment to intervene. So this book is in part an effort to explore why eco/feminism rarely figures in feminist histories, why this absence might matter, and how we might tell different stories of feminism's past, and hence its presents and futures. Thus, I seek to multiply feminism's past histories, to disrupt and refuse the dominant narratives of the recent feminist past. Geoff Bowker (2005, 230) has suggested that "if we want the future to be other than it seems to be turning out, we must create a past that is other than it seems to have turned out. People, planets and purgatory deserve multiple pasts. Only an open past can unlock the present and free the future."

Although I was very taken by the intense activism of the summer of 1993, I was also interested in what came before and what came after, with the ongoing everyday of women's lives and how this was changed by experiences at the Clayoquot Peace Camp. Difficult as it was for many to "choose" to be arrested, it was also challenging – having faced the destruction of the temperate rainforest and the tremendous and transformative life of the camp – to find ways of being in the world while that destruction continued. As Frances Beal usefully articulated the challenges of sustaining activism and revolutionary change,

> We must begin to understand that a revolution entails not only the willingness to lay our lives on the firing line and get killed. In some ways, this is an easy commitment to make. To die for the revolution is a one-shot deal; to live for the revolution means taking on the more difficult commitment of changing our day-to-day life patterns. (quoted in Springer 2005, 1)

As the women at Clayoquot knew then, or at least knew by the end of summer 1993, saving the planet, or even an apparently small part of the planet such as Clayoquot, involves more than getting arrested. The questions persist – how to sustain the revolution, how to avoid the burnout experienced by so many. How to go and on and on.

In her account of queer spiritual spaces, Sally Munt (2009, 13) recognizes that "'You can't kill the spirit' wasn't just a song, more of a creed to Greenham Women, a mantra intended to inspire and motivate." It was also an instance of "theory in its feminist travels" (King 1994), and its movements and repetitions constitute some of feminism's multiple histories and sites of emergence.

My account of Clayoquot is my effort to generate a genealogy of eco/feminism – and Clayoquot – that is recognizable to me, and I hope to others. It will become clear that this genealogy is "an unnatural history," following Terry Tempest Williams's *Refuge: An Unnatural History of Family and Place* (1991). In telling stories of Clayoquot, I attempt to make familiar stories of eco/feminism strange in order to create more open histories of eco/feminism, to write out of situated knowledges and partial perspectives, and in the process to hold open possible future feminisms and eco/feminisms – part of a story of how we go on and on and on.

Acknowledgments

I usually feel blessed that I have so many wonderful people in my life to be thankful to. However, I've learned late in the day that when it comes to writing acknowledgments, my fear of overlooking somebody crucial might have been allayed by having fewer people to thank!

First and most importantly, I would like to thank the Friends of Clayoquot Sound and all those whom I interviewed in Clayoquot, elsewhere on Vancouver Island, and in Vancouver, who invited me into their lives and homes – this book would not be possible without their considerable generosity in sharing their lives with me.

I want to thank John Othick and the Centre of Canadian Studies, Queen's University Belfast, for the initial funding to go to Clayoquot, long before I had this book in mind. Funding for the research upon which some of it is based included a graduate teaching assistantship from the University of Sussex and awards from the British Association of Canadian Studies, the AL Charitable Trust (University of Essex), and a postgraduate award from Canadian Studies at the University of Sussex. Subsequently, a Sociological Review Research Fellowship at the University of Keele, as well as time in the Centre for Research on Socio-Cultural Change (CRESC), at the University of Manchester, have contributed to the development of the book.

My initial ideas on this work were certainly shaped by my wonderful experiences in the women's studies program at Queen's University Belfast and by conversations with its students and staff, especially Jennifer Fitzgerald, and also Till Geiger and Hilary Owen. Big thanks also to Colin Coulter, Peter Doran, and Lisa Fagan, and other Belfast friends at the time, including Annie Killen, who has made a felicitous reappearance. From my time at Sussex, I thank Peter Dickens for getting this work going and Jane Cowan for generously helping me to bring it to completion. Thanks also to Sasha Roseneil and Annis May Timpson for their support and many helpful suggestions. Also from my time at Sussex, I thank Al Thomson for his work on oral history – and Dorothy Sheridan for pointing me in this direction. This book would not be what it is without my encounter with oral history in the then Centre for Continuing Education and with other students on the course, including Lorraine Sitzia and Vijay Reddy. My time at Sussex would have been less enjoyable without conversations in the feminist theory reading group (including Charlotte Adcock, Bridget Byrne, Cate Eschle, Andrea Hammel, Joan Haran, Jude Reddaway). Thanks to Chris Abuk for soup and solidarity; to Margaretta Jolly for stories of Greenham, Seneca, and for *Sisterhood and After;* and to Jenneth Parker for sharing another Sussex PhD on ecofeminism.

At CRESC, I particularly thank Medina Aitieva, Michelle Bastian, Niall Cunningham, Isabelle Darmon, Francis Dodsworth, Jill Ebrey, Gillian Evans, Jacqui Gabb, Gemma John, Hannah Knox, Fabiana Li, Shinobu Majima, Andy Miles, Johnna Montgomerie, Stewart Muir, Madeleine Reeves, Evelyn Ruppert, Mike Savage, Simone Scherger, Ebru Soytemel, Gindo Tampubolon, Daniel Tischer, Mike Upton, Farida Vis and Vron Ware, who have shaped my thinking, and provided company over the past years.

For ongoing and prolonged discussions, critical commentary on the text, and continuing support, I am grateful to Bridget Byrne, especially for early reading of transcripts; to Joan Haran, for reading and re-reading and re-reading, for all the many, many e-mail and face-to-face conversations, and for passionate engagements with feminism, eco/feminism, and *The Fifth Sacred Thing;* to Anne Rudolph for transcription and endless support in a myriad of forms; to all the Hambles for living with this for many years; to Carol and Sue for being there at Clayoquot and after; to Dawn for helping to make the book real; huge thanks to Jeannette for more than I can ever imagine.

And to friends beyond academia, who lure me away from the desk sometimes and allow me to feed them (and in return allow me to be fed in multiple ways): Rebecca J., Rebecca T., Sally, Thas, Theres, and in Brighton, Chris for being there since Fresher's Day and the Women's Group, to Kat and Scrap more recently, Phillip for the haircuts and chats, Rebecca D. for sharing laundry and poems.

And I remain, as I hope will be clear, inspired and indebted to all those who have written on eco/feminism before me and whose work has made this book possible.

I am particularly grateful to the two reviewers of the manuscript for their enthusiastic engagement with previous drafts, and constructive and careful, and caring, suggestions. I could not have wished for better readers. Thanks also to Kate McNicholas Smith for timely and attentive indexing.

And there are many not named here, but not forgotten either, who are also in my thanks for providing advice, support, and encouragement along the way. There are many who will have heard too much about the book for too long. I appreciate their patience, forbearance, and continuing interest in this work.

I thank Jools for allotmenting, for balance, for frolicking in the sea, for scrabbling – also with the book! – and for the hope and possibility of one day returning the favours.

And this book is dedicated to Amelia, Claire, and all of the overlapping Young Women's Group and allotment project, LGBT Youth North West, Feminist Webs, and the Sidney Street Cafe – to Ali, Alison, Damien, Felix, H., Heather, Janet, Kelly, Lily, Liz, Maggie, Myrtle, Rachel, Sahar, Sally MBE, Sophie, Tash, and all, all of whom matter, and matter hugely, and for the sheer pleasure of getting to grow together down the allotment and through Feminist Webs.

Sections from this book were previously published as:

N. Moore, "Ecofeminism, Non-Violence and the Future of Feminism," *International Feminist Journal of Politics* 10, 3 (2008): 282-98.

N. Moore, "Eco/feminism and Rewriting the Ending of Feminism; From the Chipko Movement to Clayoquot Sound," *Feminist Theory* 12, 1 (2011): 3-21.

Shorter extracts appeared in:

N. Moore, "Imagining Feminist Futures: The Third Wave, Postfeminism and Eco/feminism," in *Third Wave Feminism: A Critical Exploration*, 2nd ed., ed. S. Gillis, G. Howie, and B. Munford (London: Palgrave Macmillan, 2007), 125-41.

As well as the first edition:

N. Moore, "Ecofeminism as Third Wave Feminism? Essentialism, Activism and the Academy," in *Third Wave Feminism: A Critical Exploration*, ed. S. Gillis, G. Howie, and B. Munford (London: Palgrave Macmillan, 2004), 227-39.

And also in:

N. Moore, "Ecological Citizens or Ecoterrorists? Learning through Environmental Activism in Clayoquot Sound," in *Adult Learning, Citizenship and Community Voices: Exploring Community-Based Practice*, ed. P. Coare and R. Johnston (Leicester, UK: National Institute of Adult Continuing Education, 2003), 92-107.

THE CHANGING NATURE OF ECO/FEMINISM

1

Rethinking Eco/Feminism through Clayoquot Sound

Eco/Feminism and Clayoquot Sound

In the early 1990s, as accounts of the end of feminism proliferated and gathered pace, I encountered a vibrant eco/feminist politics and activism in Clayoquot Sound, on the west coast of Canada. In the summer of 1993, activists set up a peace camp at Clayoquot (pronounced "klak-wat") to support the blockading of a road into an extensive area of temperate rainforest that was slated for clear-cut logging. Throughout the summer, more than eight hundred people were arrested for non-violent civil disobedience while blockading the logging road. The arrests were followed by the "Clayoquot mass trials," where all defendants were found guilty of criminal contempt of court. Many served jail sentences or spent time on electronic monitoring; fines were commonly set at $1,000. The protests brought international attention to a local environmental group – the Friends of Clayoquot Sound (FOCS) – and to the clear-cutting of Canada's temperate rainforest. Clayoquot has most visibly been a locus of environmental controversy, but to understand events there as being solely about environmental politics is to restrict oneself unnecessarily to a limited notion of the political, an impoverished notion of the environmental, and a narrow view of (the nature of) the "nature" that is always so bound up with environmental politics.[1] The Clayoquot Peace Camp was said to be based on feminist principles, which were sometimes explicitly identified as *eco/feminist.* Indeed, its success has often been attributed to this eco/feminist ethos. The paradox for me was encountering this dynamic eco/feminist politics at a time when the end of feminism was widely being declared. I have puzzled over this apparent disjuncture ever since.

The importance of eco/feminist politics was picked up beyond the camp and sparked the popular imagination. The *Vancouver Sun* even published an article titled "Eco-Feminists Run 'Peace Camp' at Clayoquot Sound" (S. Bell 1993) – although the notion that they were "running" the camp somewhat misconstrued their eco/feminist practices. For many, the eco/feminist dimensions of the campaign were compellingly conveyed in the award-winning documentary film *Fury for the Sound: The Women at Clayoquot,* directed and produced by Shelley Wine (1997), who had spent time at the camp. Shown on Canadian television, *Fury for the Sound* brought many of the women involved in the Clayoquot campaign into the living rooms of those who never made it to the camp. Yet, although Clayoquot has garnered some attention in academic writing on environmental politics, the eco/feminist dimensions of the campaign remain largely unexplored, except for a few unpublished dissertations.[2] In the absence of sustained academic attention, some of those involved have produced their own accounts of the campaign. For example, Jean McLaren recorded life at the camp in *Spirits Rising: The Story of the Clayoquot Peace Camp* (1994); Betty Krawczyk, one of three grandmothers jailed on the first day of arrests for refusing to sign an undertaking not to return to the blockade, wrote autobiographically in *Clayoquot: The Sound of My Heart* (1996); and Chris Lowther's *A Cabin in Clayoquot* (1997) recounted her daily life on the west coast as well her experiences of the summer of 1993 (published and distributed locally, the works by Krawczyk and Lowther had a considerable readership on Vancouver Island and in Vancouver). In "Women Out Front in Clayoquot Sound," Valerie Langer and Jan Bate (1993) reflected on women's activism in the campaign. In his fictionalized *The Garden Club and the Kumquat Campaign,* Des Kennedy (1996) wrote thoughtfully of women and feminism, as did Bob Bossin in "The Clayoquot Women" (1999).

None of these works offers an easy, or doctrinaire, definition of what it means to say that the camp was eco/feminist, or informed by eco/feminism. Describing the complexity of Clayoquot politics is no simple or straightforward task. Events in Clayoquot exceed normative understandings of both gender relations and feminist politics, *including* many feminist and eco/feminist theorizations of these. For many, deforestation was not obviously a women's, or a feminist, issue, in the way that, say, childcare or reproductive justice might be, and though the camp was described as feminist and eco/feminist, it was not women-only. Neither can the campaign be reduced to common accounts of eco/feminism. It was not simply the case that women showed the strength of their commitment to caring for children, or even for the whole planet and its future, by risking arrest and spending time in jail. Neither was it quite the case that they exceeded traditional roles as wives and mothers by taking to protest and blockades – an explanation frequently offered for women's environmental activism, but one that risks constraining what is at stake in such campaigns. Rather, I suggest that gender relations and feminist politics at the camp and throughout the campaign were altogether more complex and entangled. Thus, the circulation of debates about feminism and the articulation by some of a specifically *eco/*feminist politics deserve further consideration, albeit this is a complicated task. Indeed, the slippage between the terms "feminism" and "eco/feminism" already suggests some of the latent, and not so latent, tensions that are the subject of this book.

I signal these tensions through my use of the term "eco/feminism" to gesture toward the sometimes fruitful, sometimes unproductive, relationship between eco/feminism and feminism. Eco/feminism is both "of feminism" and simultaneously offers a critique of it. This is not unusual, as feminism has often been defined by such tensions. There has been a proliferation of different feminisms as critiques have been made by lesbian feminists, black feminists, and the disability movement, to name a few of the most salient. I hold on to this label of eco/feminism as productive at this juncture in feminism, to signal a specific constellation of interests that cannot be assumed under the rubric of "feminism" alone (Moore 2004).

The Clayoquot Peace Camp and accompanying blockades emerged at a difficult moment for many feminists. Far-reaching questions about the possibility of feminist politics were being raised, and the category of "woman" as the very subject of feminism was being called into question. Certain foundational tenets of feminism, such as a reliance on women's experiences for theory building, were being exposed as implicated in exclusionary practices. These criticisms called into question attempts to ground feminist politics in the category of "woman" and to link women's experiences, of oppression or of activism and resistance, through any simplistic notion of global sisterhood. The proliferation of feminisms through the 1980s and 1990s, including eco/feminism, had the effect of creating a sense of uncertainty about the future of feminism. However, the most damning accusations of essentialism and universalism were manifest in assertions that dominant accounts of feminism marginalized black and Third World feminisms, and lacked a thorough understanding of the implications of histories of colonialism and racism for feminism itself, thus demonstrating that feminism relied on its own exclusions. This painful realization of the non-innocence of feminism and the challenge of dealing with the reverberations of these trenchant critiques continue to echo through feminism today.

Attempts to comprehend the politics of Clayoquot founder for many reasons: the challenges of articulating eco/feminist politics; the tense state of feminist politics at that time and since; and specifically, the troubled place of essentialism and universalism in these debates. The identification of any political or theoretical position as essentialist has characterized many problematic struggles among feminists, and the label "essentialism" has often been applied to eco/feminism. In this context, eco/feminism is merely another item on a long list of conflicts over essentialism, which includes overlapping debates regarding maternalism, feminist peace politics, the "sex wars," and the work of Carol Gilligan and Sara Ruddick. Eco/feminism has been particularly vulnerable to critique because it is so easily caricatured due to its explicit engagement with "women" and "nature." It has been rendered suspect by concerns that it presents women as being closer to nature than men are and especially by claims that link this with women's role as childbearers and childrearers, something that feminists had worked hard to undermine. In seeking to re-open the apparently closed question of women and nature, eco/feminists have had difficulty in convincing other feminists of their feminism, particularly given reservations about maternalist and universalizing discourses (see, for example, Sturgeon 1997, 167-68). Because "essentialism" has become such a pejorative term, eco/feminists have often found themselves in the frustrating position of seeing their arguments dismissed and of not being perceived as proper feminists themselves. Many eco/feminist academics responded to this with an insistent anti-essentialism, or in Noël Sturgeon's case, an argument for strategic essentialism. Many have also engaged in a disavowal of apparently essentialist strands within eco/feminism itself.

Yet there was little evidence of essentialism at the Clayoquot camp. Indeed, essentialism seemed to offer a very limited way of understanding the politics of Clayoquot. As Sturgeon (ibid.) has argued, essentialism has been produced as one of the key concepts on which certain feminisms, particularly activist eco/feminisms, are seen to founder. For me, Clayoquot offers an exemplary site where certain significant debates in feminism have been productively opened up; moreover, eco/feminism is a particularly useful focus because of its willingness to engage with feminism's abject other, "nature." The particular imbrications of feminist and environmental (and other) politics at Clayoquot render it a fascinating site for an examination of feminism. These intersections, of feminism and environmentalism, are particularly interesting because hegemonic notions of the "natural" underlie many of the justifications for women's exclusion from the public sphere, from politics and women's confinement to the home, to domesticity, to the private, to passivity. This book argues that this nexus of concerns is a fruitful site for the examination of political agency, activist subjectivities, academic theory, and the complex role of "nature" in these debates.

Clayoquot offered me a potent opportunity for exploring a whole range of issues of ongoing concern for feminists. In 1993, as I returned from Canada, I was excited by the vibrancy of the campaign, but at the same time I was challenged by the difficulty of bringing what was happening in Clayoquot into conversation with mainstream feminism. The sheer energy of eco/feminist politics at the camp appeared to confound dominant narratives – that the women's movement was dead, that feminist activism had waned throughout the 1980s and early 1990s, that feminist politics were no longer possible. There was a stark contrast between what I had experienced at Clayoquot and the prevailing narratives of the end – or death – of feminism.

One of my ongoing fascinations is, therefore, the paradox of the vitality of eco/feminist activism, with its insistence on transnational connections, from the Chipko Movement in India, to the Kenyan Greenbelt Movement, to Greenham Common in the United Kingdom, in contrast to the widely circulating accounts of the end of feminism and especially of global sisterhood, in the 1990s and since. Despite the pervasiveness of feminism's hegemonic narrative, others have also recognized this contradiction – for instance, Mary Hawkesworth (2004, 962) observed that "a strange phenomenon has accompanied the unprecedented growth of feminist activism around the globe: the recurrent pronouncement of feminism's death." Hawkesworth (ibid., 983) pointedly noted that "with no corpse, no proof of demise, just vague hints of self-inflicted wounds and natural causes, feminism's death by report erases the social justice activism of women around the world while covering the traces of the erasure." In an effort to produce versions of feminism that are better than others, feminists have prematurely terminated a variety of feminisms, specifically eco/feminism and activist feminisms. The grounds for termination are usually lack of theoretical sophistication and, in particular, manifestations of essentialism, charges of which activists are often found most guilty. Hawkesworth's (ibid.) suggestion that "feminists might want to target this cultural terrain [reports of the death of feminism/the proliferation of feminist activism] for intensive action" is a challenge that this book takes up, offering one such engagement with debates about the death of feminism and the erasure of feminist activism, specifically of eco/feminist activism.

Although an eco/feminist peace camp may seem an archetypal site for the reinscription and repetition of essentialism, I suggest that without returning to such sites, we cannot productively work through feminist dilemmas about the nature of nature. This book turns to Clayoquot, to activism in the early 1990s, as an exemplary site through which to engage with key conversations in eco/feminism and its recent history. Through a skeptical reading of accounts of essentialism, I have come to understand this late-twentieth-century peace camp, not as a quaint throwback to the disavowed activism of the 1970s and 1980s, but as a site through which the future of eco/feminist politics was, and can be, (re)imagined.

Although the peace camp and blockades transformed the terrain of environmental and particularly forest politics in Canada and beyond, and Clayoquot retains a prominent place in Canadian and international *environmental* political histories, Clayoquot does not – yet – occupy an analogous place in national or international feminist or eco/feminist imaginaries. In fact, the same can be said of eco/feminism itself, which does not often figure in overviews of feminism or in the recently proliferating histories of it. To understand the curious state of affairs whereby (some) academics have bemoaned the dearth of feminist action even as eco/feminist activists were busily engaged in campaigning across the world, it is first necessary to understand how and why essentialism has come to dominate, and constrain, discussions of eco/feminism. I begin this exploration by focusing on its two most commonly identified manifestations: maternalism and the universalizing implications of linking women and nature.

The Challenge of Essentialism: Maternalism and Universalism

Although the term "essentialism" may have little meaning for activists, it is redolent with significance for feminist academics. Challenging essentialism, and especially biological determinism, has long been a central project of feminism and a focus of much feminist theory. A founding tenet of second-wave feminism was to dispute arguments that women's nature was to nurture. Such claims were held to be essentialist, defined by Diane Fuss (1989, xi) as "a belief in the real, true essence of things, the invariable and fixed properties which define the 'whatness' of a given entity." For many feminists, the reliance on biology to establish women's essence has been a focus of critique. The biological qualities usually attributed to women justified their confinement and innate suitability for the private, domestic sphere, as well as their exclusion from education and politics. Thus, undermining associations between women and nature became a key project for feminists, who drew on Simone de Beauvoir's (1953, 249) famous dictum that "one is not born, but becomes a woman" to challenge biological determinism. Distinguishing between sex and gender, feminists argued that though sex was biological, gender and gender roles were socially constructed.

The distinction between sex and gender, the biological and the social, was crucial in suggesting the possibility of change. Whereas biology and nature were perceived as fixed and unalterable, the social was felt to be mutable – hence the importance of the term "gender" for feminists in holding out the prospect of social transformation. Challenging assumptions about the nature of women and insisting that women's oppression was a political issue were crucial steps for feminists. Countering biological determinism and other essentialisms subsequently became an important political strategy for feminists. Thus, Christine Delphy (1984, 144) summed this up as

> people do not revolt against what is natural, therefore inevitable; or inevitable therefore natural. Since what is resistible is not inevitable; what is not inevitable could be otherwise – it is arbitrary therefore social. The logical and necessary implication of women's revolt, like all revolts, is that the situation can be changed. Belief in the possibility of change implies belief in the social origins of the situation.

However, though accounts of the possibility of change relied on the notion of gender as a social construct, they equally depended on an assumption of the fixity of nature and of the biological (I return to the significance of this in Chapter 9).

The rejection of essentialism has been key for feminism, but arguably essentialism ultimately threatened to undermine feminist politics. Although the critique of essentialism initially defined the "outside" of feminism, the non-feminist, the anti-feminist, and patriarchy as such, it was eventually applied "within" feminism itself. Hence its very pejorative tone and the implication that eco/feminists were not proper feminists at all. Furthermore, there is only a short slide from "not a proper feminist" to non-feminist or anti-feminist, and attempts were made to resolve the dilemma of essentialist feminisms by relegating them to a naive past. Perhaps more precisely, the recognition that essentialism existed within feminism revealed the fiction of the distinction between "inside" and "outside," between a feminism separate and apart from the non-feminist (see, for example, de Lauretis 1990, 268n7).

In explaining the predominance of women in environmental activism, Elizabeth May, former Sierra Club of Canada director and current leader of the Green Party of Canada, voiced the kind of argument that is commonly read as essentialist:

> Everyone had their own, different answers ... And then there was my answer for which I was roundly criticized by politically correct feminists. My answer was and still remains that women are essentially different than men. We operate more from an intuitive thought process. We are biologically and spiritually connected to the cosmos, its planetary shifts, the earth's tides and phases of the moon. We are more nurturing, more concerned with the flow and flux of life-people, plants, animals, even seas. Consciously or not we find ourselves part of the Gaia, part of the living planetary whole. (quoted in Felesky 2001)

Such accounts have proved contentious. Canadian eco/feminist Catriona Sandilands curtly summed them up in *The Good-Natured Feminist: Ecofeminism and the Quest for Democracy* (1999, 5): "The fact of being a woman is understood to lie at the base of one's experience of ecological degradation; of one's interests in ecological protection, preservation, reconstruction; and of one's special ecological consciousness." Thus, many eco/feminists distanced themselves from essentialism, as I explore in Chapter 2, insisting on eco/feminism's potential to challenge and problematize, rather than reify, maternalist and other essentialist representations of women and nature. Sandilands (ibid., xiii), for instance, opened her book with a sense of alarm about current portrayals of women's environmentalism: "Women's concerns about nature, even if they have eventual public appearance and impact, boil down to an obvious manifestation of natural protective instincts towards home and family." As Sandilands (ibid., iii) put it, "The neoconservative aroma of this discourse should be quite noticeable: a return to patriarchal and heterosexual 'family values' will restore not only a healthy (natural) family but a healthy (natural) planet."

An equally challenging critique of eco/feminism's political credibility has been that it universalizes women-nature connections and homogenizes differences between women. The Chipko Movement, in which women campaigned to prevent deforestation in India by hugging trees ("chipko" means to hug or embrace), has proved a powerful image for eco/feminists. The ritual citation of Chipko as the iconic instance of eco/feminist activism was particularly prevalent throughout the late 1980s and early 1990s, when eco/feminists linked diverse activisms across the world under the banner of eco/feminism, from Chipko to the Kenyan Greenbelt Movement to Greenham Common. Yet this attempt to connect diverse movements has been seen as redolent of eco/feminism's ongoing struggle, and failure, to negotiate the challenges of race and difference. The lack of substantive accounts of what it meant to name Chipko as eco/feminist has left eco/feminism open to criticisms of reproducing stereotypes of Indian women as either victims or saviours of the world, and of universalizing and essentializing relationships between women and nature. This was picked up in particular by those in the field of development studies who had conducted extensive in-depth ethnographic fieldwork, which presented nuanced and situated accounts of women's relationships with nature. Such writers identified eco/feminism as a Western phenomenon that could not easily be translated to, or imposed on, non-Western contexts, and they resisted reading Chipko as an instance of eco/feminist activism, preferring to understand it in the context of a history of peasant resistance in the Garwhal Himalayas. The elision of difference by (Western/Northern) eco/feminists was seen by many as a universalizing strategy and yet another manifestation of essentialism; hence, it provided ample justification for dismissing eco/feminism.

Given these critiques, which lie at the centre of some of feminism's most fraught debates, it is not surprising that eco/feminism may appear anachronistic. Yet despite these trenchant criticisms, one of my aims in this book is to articulate why eco/feminism remains a crucial intervention in particular feminist debates. Many eco/feminist theorists have responded to criticisms of essentialism by producing sophisticated accounts of eco/feminism's anti-essentialism, but it is not clear that this strategy has succeeded. In my view, the anxiety over essentialism has placed eco/feminism in an unnecessary stranglehold from which it has struggled to escape. My project differs in that it ambitiously seeks a route out of the tripartite bind of essentialism, anti-essentialism, and strategic essentialism.

From the Chipko Movement to Clayoquot Sound

Misha Kavka (2001, xii, emphasis in original) wrote, "If 'feminism,' then, begs definition and yet refuses to be singly defined, one approach is to take stock of where (rather than *what*) 'feminism' is by looking back over where it has been – to consider, in other words, the work of today as the consequences of the histories accruing in the last thirty years in the name of feminism." In my attention to a manifestation of eco/feminist politics in a particular place and time – Clayoquot in 1993 – I may appear to take this *where* rather more literally than Kavka intended. My intent in doing so is to counter the tendency of some feminists to reiterate the importance of a politics of location and situated knowledges while glossing over troubling questions of (colonial) histories, geographies, and ecologies, and the complex travels and translations of feminism. My turn from the Chipko Movement and other instances of eco/feminist activism in the South to examine women's environmentalism in the North is intentional. Central to my reading of women's environmental activism is attention to the specificities of its histories, to the particularity of nature in specific contexts (including national ones), as well as to exploring practices of mothering and the invocation of maternalist discourses, and to unpacking eco/feminism's claims of international grassroots activism. All of this is necessary in any effort to recuperate eco/feminism in the fraught climate created by critiques of essentialism.

The temperate rainforest of Clayoquot, which constitutes "nature" in the Canadian context, carries an extraordinary symbolic power. The conflict was most visibly fought over the impact of clear-cut logging, and it gained public currency as a struggle over jobs versus the environment, but it also touched on issues fundamental to many articulations of Canadian identity: nature, particularly nature as wilderness. In protesting clear-cutting, activists were also engaging in a tense contestation over the meaning of being Canadian, which had the effect of bringing them into conflict with the Royal Canadian Mounted Police (RCMP) and the courts. Whereas Canada is highly invested in the idea of its forests, many of its politicians and business people are committed to logging and the economic expansion that it facilitates. Ironically, the activists who tried to protect the forests from clear-cutting were often perceived as traitors.

For me, stories from Clayoquot provide a salutary contrast to dominant narratives of nature as wilderness. In *Survival: A Thematic Guide to Canadian Literature,* Margaret Atwood's (1972) classic effort to define a national literature, nature and wilderness were overwhelming and threatening places for humans. For Atwood (ibid., 33), Canadian stories were accounts of "hanging on" or "staying alive," of those who "made it back from the awful experience – the North, the snowstorm, the sinking ship – that killed everyone else." She (ibid., 35) wrote that "given a choice of the negative or positive aspects of any symbol – sea as life-giving Mother, sea as what your ship goes down in; tree as symbol of growth, tree as what falls on your head – Canadians show a marked preference for the negative." In *Survival,* Atwood was more explicitly concerned with the survival of humans than with that of nature. Later she (1995, 115-16) revisited and revised her survival thesis:

> Next time you're in Pearson International airport in Toronto, wander through the souvenir shops and look at what is to be found there in the way of instant national identity. Chances are you will find a light-up Mountie or two, some maple leaves made out of maple sugar, many notecards with birds, animals or landscapes on them, some Eskimo carvings, some smoked salmon in wooden boxes with Haida motifs, and a lot of mittens. You will also find many gift books with names like *Beautiful Canada,* and these will have a large supply of Northern scenes – scenes shown as vast, empty, untouched, luminous, numinous, pristine, and endless. Canadians have for long taken the North for granted, and we've invested a large percentage of our feelings about identity and belonging in it.
>
> But the bad news is coming in: the North is not endless. It is not vast and strong, and capable of devouring and digesting all the human dirt thrown its way. The holes in the ozone layer are getting bigger every year; the forest, when you fly over it in a plane, shows enormous wastelands of stumps; erosion, pollution, and ruthless exploitation are taking their toll ...
>
> The North will be neither female nor male, neither fearful nor health-giving, because it will be dead. The earth, like trees, dies from the top down. The things that are killing the North will kill, if left unchecked, everything else.

Although Atwood was talking specifically about how the North figures in Canadian imaginaries, her reflections resonate more widely with accounts of wilderness as central to Canadian identity. In updating her original survival thesis with the recognition that wilderness itself may not endure, her ultimate theme remains one of human survival. The threat no longer comes from bears and winter storms, but from climate change and deforestation – from the environmental destruction wrought by "man's" entry into nature/wilderness. Atwood's survival thesis can be seen as flawed, as I have argued elsewhere (Moore 2003, 99):

> The notion of the Canadian-ness of wilderness obscures the fact that the emergence of the Canadian nation was predicated on the exploitation of "natural resources," on the colonisation of "empty" landscapes, and on genocide. At the same time as the Canadian government promotes the representation of "wilderness" in Canadian art and literature, and its consumption in National Parks, it customarily licenses its destruction.

This contradiction between the reification of wilderness and the destruction of nature and of those associated with it is a central tension at the heart of the Canadian state. What has been omitted from Atwood's account is that the hacking out of Canadian identity from the backwoods and the very creation of the Canadian nation-state relied on the destruction of the trees and wilderness creatures, and crucially, of the First Nations, who do not figure in either her earlier or later accounts, except as Eskimos and Haida motifs. Atwood concluded by constructing "everything else" – white Canadians? – as the potential victim of this "death of nature." In a perverse circumlocution, she made a small shift from her original survival thesis, from Canadians as victims of nature's overwhelming power to Canadians as victims of nature's vulnerability. Notwithstanding her public support for Clayoquot, Atwood relied on glossing over how Canada, and Canadians, might be implicated in national and global environmental degradation through the logging industry, international trade, and domestic consumption practices.

Atwood's lack of attention to unpacking "nature" is suggestive of a significant investment in liberal versions of the nation, which persisted even when she discussed the subject of women and wilderness. Recognizing that the dominant stories of Northern landscapes were those of "man" and wilderness, she asked what happened when women entered Northern landscapes that were traditionally occupied by men. Atwood (1995, 95-96, 101, emphasis in original) identified two waves of women settlers; the first consisted of the early nineteenth century settlers who came with their husbands. The second wave, in the twentieth century, came for different reasons:

> Instead of going off into the woods to be with a man, they start going off into the woods to be by themselves. And sometimes they are even doing it to get *away* from a man. A good many women's novels have been written in response to the theoretical question *Where did Nora go when she walked out of the doll's house?* but the solutions have varied according to locale ... But one of the answers that has frequently presented itself to Canadian women writers has been "off to the woods," or, at the very least, "off to the summer cottage."

So, in Atwood's world, women don't storm off to the North; they go to the woods or to the summer cottage, suggestive of altogether gentler landscapes. Atwood did not go so far as to explicitly suggest that *women's survival* may be predicated on the existence of woods to which they can escape and that would result in a rather different survival thesis. Neither did she pursue the question of what women did once they stomped off to the woods. Marnie Andersen began to answer this question in *Women of the West Coast: Then and Now* (1993), although, like Atwood, she did not explore the stories of women who lived in Clayoquot before "pioneers" and "settlers" arrived. Initially, she intended to write about women pioneers, settlers, and homesteaders, but her focus shifted to environmentalism and women environmentalists. Andersen (ibid., x) reflected,

> My older friends took me – in my imagination – back to the earlier, gentler, but no less challenging days of settling the west coast. When the biggest excitement of their lives was, depending on their age (and marital status), the arrival of the *Princess Maquinna* – their only link with the outside world – or the weekly dance in the community hall. When an outing to the general store meant a six hour hike (round trip) through dense rainforest, or a trip down the inside passage by rowboat.
>
> Today the west coast woman often arrives by van. She is likely to be jolted by the west coast experience while paddling a kayak, hiking to the Meares Island nature trails, or observing whales feeding in the wild. However it happens – to any one of us – the experience is unforgettable, and often requires a complete change in lifestyle. Some go home, pack up their belongings, and migrate west, others simply reassess their values and priorities, while still others sit down and write about it.

And some come, and stay, and become environmental campaigners. So who were the women of Clayoquot who set up the Peace Camp and who came and were arrested, and what were they doing? Were they getting back to nature? Were they reproducing the heterosexual nuclear family while saving the planet? Were they reproducing colonial and imperial natures? Were they becoming earth mothers? Were they being essentialist or even strategically essentialist? Were they universalizing women-nature relationships?

This raises a series of other questions, which seem much more interesting and pertinent to me: How adequate, or even useful, are these questions for understanding the lives and actions of the women at Clayoquot and their relevance for feminist politics? How have they come to dominate the stories of eco/feminism? How has eco/feminism become so entangled in these stories? How might we ask other questions of eco/feminism, and of women and nature, and how might we – I – tell different stories of feminism, eco/feminism, and women and nature?

Approaching these questions through Clayoquot Sound and Vancouver Island is instructive. The history of Vancouver Island, which is also a history of colonialism, is but one salutary reminder that women have not always been associated with nature and that nature is not always seen as gentle. In colonial projects, bourgeois white women have often served as civilizing influences, sent to colonies to undertake teaching and mission work for both indigenous people and the white men who laboured in the mining and logging industries. In this instance, it was white men who were in danger of being too close to a savage nature, in the homosocial life of the logging camps or in going Native by entering into relationships with First Nations women (Ware 1992; McClintock 1995; Perry 2001). Though Vancouver Island was only briefly a colony (1849-71) before it entered Confederation, I take up Adele Perry's argument in *On the Edge of Empire* (2001, 6) that "this does not imply that the relationship of colonialism existed only within the tidy parameters of these dates, but simply indicates a specific moment in state formation." I follow Perry and others who understand British Columbian history (and Canadian history more generally) as colonial history. So, despite Andersen's invocation of "earlier, gentler" days, the colonization of the west coast was certainly not gentle for everyone, not, most obviously, for the First Nations who already lived in Clayoquot, for the animals who bore the brunt of the fur trade, or for what was then densely forested land.

Despite Atwood's account, not all the first-wave white women were accompanied by husbands. Some were sent from Britain on bride ships, fated to become wives once they arrived. Sending respectable white brides from England to Victoria was just one of the reform efforts that endeavoured to transform island logging camps into a white settler colony and that followed the common path of considering bourgeois women as agents of empire (Ware 1992; McClintock 1995). As Perry (2001, 3) recorded, in BC, First Nations people significantly outnumbered whites for a considerable time, and men outnumbered women in the small European community. Thus, homosociality in backwoods logging camps and the phenomenon of white loggers establishing relationships with First Nations women threatened efforts to establish British Columbia as "an orderly, respectable white settler colony." Between 1849 and 1871, four "assisted immigration projects" (bride ships) were promoted in Britain and supported by middle-class feminists who were seeking ways to promote women's independence and to address popular opposition to women's waged work. The Female Middle-Class Emigration Society, set up in London in 1862, aimed to address the problem of "surplus" women in Britain and presented assisted women's emigration as a route to preserving middle-class respectability. Yet, as it became clear that colonial Victoria was not entirely suitable for middle-class women, the society ultimately focused on working-class women, and on young women, as potential wives for working men, as servants for the emerging middle classes, and as governesses for their children. Perry's (2001, 166) careful history described "working class women's failure to behave in ways consistent with colonial discourse," as well as a persistent indigenous opposition to colonialism. Thus, Perry (ibid., 4) found that "British Columbia's colonial project was a fragile one that was constantly challenged both by First Nations resistance and by white unwillingness to conform to prevailing constructions of appropriate behaviour and identity." It is useful to bear these histories in mind if we are to understand the women who came to Clayoquot much later, as well as the Clayoquot Peace Camp and its disruptive effects.

These histories are key for understanding the importance of feminism, and eco/feminism specifically, in the Clayoquot campaign. This can be illustrated through two very different accounts of Clayoquot, which I use to demonstrate the importance of fully considering the critical difference that eco/feminism made in the campaign. In their collection *A Political Space*, Warren Magnusson and Karena Shaw (2003b) presented Clayoquot as offering a puzzle. For Magnusson "the puzzle of the political is especially apparent" in Clayoquot, which led him to pose the questions: "What forms does politics take? How are we to relate to it?" He reflected that "In

Clayoquot, the appropriate responses to such questions are unclear, and the answers we get from the conventions of political science are not very helpful" (Magnusson 2003, 1). A key aspect of the puzzle of Clayoquot is the challenge of locating Clayoquot. Magnusson and Shaw note that Clayoquot "appears to be at the *periphery* of contemporary power and authority and thus to be *marginal* to the study of politics" (Magnusson and Shaw 2003c, vii, emphasis in original). But they go on to argue that the apparent marginality of Clayoquot "is an effect of particular assumptions, assumptions that need to be challenged. Clayoquot is more usefully interpreted or 'read' as a center of interaction among the movements, powers, and authorities that produce the world in which we live" (Magnusson and Shaw 2003c, vii). In posing Clayoquot as a puzzle, they do not offer any particular theoretical framework to resolve it, but rather they propose a method which privileges the site itself, and that is "to read the global *through* the local," with the aim that "by exploring Clayoquot as a microcosm of global politics – we hope to disrupt the assumptions that constrain our political imagination" (Magnusson and Shaw 2003c, vii). The book is a thought-provoking collection of essays from a range of intense conversations at an International Workshop on the Politics of Clayoquot Sound, including academics, activists, locals, and visitors who came to Clayoquot and engaged with the site and with an extensive archive of documentation related to Clayoquot which they assembled for the workshop.[3]

In contrast, Bruce Braun's *The Intemperate Rainforest: Nature, Culture and Power on Canada's West Coast* (2002) had no problem locating Clayoquot. Braun offered a clear diagnosis of what was wrong with the campaign, and identified his own resolution to these problems. While Braun (2002, 2) insisted that he was sympathetic to protesters, whose passion and commitment deserved respect, he had concerns that "in the popular press, and in the rhetoric used by key actors, debate over the future of these forests was often cast in terms of a binary logic (pristine nature/ destructive humanity)." For Braun this binary logic authorized "certain actors to speak for nature's defense or its management," and curiously to me anyway, resulted in him aligning environmentalists *with* transnational capital and the state, and *against* those he saw as marginalized – local communities, forest workers, and First Nations. In particular Braun took environmentalists to task for perpetuating the absence of First Nations in the media framing of the controversy as one of environmentalists versus loggers. His solution then was to call for a radical postcolonial environmentalism, which for him was a politics where "wilderness" is not reified and where First Nations were made visible. Braun's method was to take a number of events, artefacts, or images, drawn if not always directly from the broad campaign, then from the broader production of natures on the west coast. His artefacts included photos of forests before and after clear-cutting and maps showing decreases in levels of old-growth forest over time. He also chose artefacts more loosely connected with what we might term a wilderness imaginary – from wilderness tourism to the art of Lawrence Paul Yuxweluptun and Emily Carr. Braun's method of reading was to juxtapose artefacts, making "no attempt to resolve them into a totality" (2002, 23); although arguably he nonetheless produced a totalizing solution – radical postcolonial environmentalism – on the basis of his readings. And it is not clear that Braun's artefacts can bear the weight of his argument.

In teasing apart the difference in these two accounts, I find the work of Anna Tsing (2005) useful, not least because of its focus on the destruction of forests, in her case in Indonesia. In *Friction: An Ethnography of Global Connection* one of Tsing's interests is collaborations between environmentalists and indigenous people, and obstructions to such collaboration. Her argument, therefore, diverged from those who read environmentalists' interest in indigenous knowledge "only as a repetition of indigenous fantasies and imperial histories." She commented that such accounts "offer a historical metanarrative in which nothing good can happen – good or bad – but more of the same. Familiar heroes and villains are again arrayed on the same battlefield. It is difficult to see how new actors and arguments might ever emerge." Tsing's (ibid., 4) use of "friction" to signal not a repetition of a "clash of cultures," but to gesture to "the awkward, unequal, unstable, and creative qualities of interconnections across difference" is a welcome contribution to efforts to rethink connection and solidarity. I too am interested in tracing situated knowledges and partial perspectives, and not repeating the same old stories about environmentalists and the reification of nature, or indigenous peoples – not least because, contra Braun, this seems to me precisely what the Friends of Clayoquot Sound were involved in. I don't want to deny that wilderness imaginaries of the kind that Braun had in mind did not emerge in and through the campaign. Indeed, it would be a surprise if they did not, given the dominance of understandings of forest as wilderness and its circulation as wilderness in industries such as tourism and even logging, as well as by provincial and national governments. Magnusson and Shaw (2003a, 272-73) note this too: "Clayoquot's image as pristine wilderness was used by environmentalists to mobilize support outside the region. Insofar as environmental activism tended to set Clayoquot aside as a special place invested with cultural meanings derived from 'nature' or 'wilderness,' it remained complicit in some of the practices it sought to overcome." Yet, as they clearly recognized, what makes Clayoquot so interesting is that there was so much more going on than a repetition of a desire for wilderness. Rather, what becomes relevant is the extent to which local activists "attempted to break out of this frame and re-pose the issues inclusively" (Magnusson and Shaw 2003, 271).

Whereas Magnusson and Shaw's approach is to read the global through Clayoquot Sound, Braun's (2002, 69) different reading sees environmentalists turning to the global. Rather than the political in Clayoquot offering a puzzle for Braun, this is proof that "the protest's leaders were savvy political actors who knew that images travelled widely and instantaneously in late-twentieth century mediascapes, and that political identifications were not necessarily contained solely within the boundaries of place, or the nation state." Or else he understands it as a displacement of local ecological issues into global arenas, facilitated by the circulation of the kinds of images he examines (ibid., 304n7). In part, this may be because he relies more on the work of organizations such as Greenpeace and the Wilderness Committee than on that of the Friends of Clayoquot Sound (FOCS), which is my focus, as well as that of Magnusson and Shaw.[4]

Throughout this book, I suggest that there was a lot more at stake in the turn to the international than the tactics of media-savvy protesters. For instance, William Chaloupka's chapter in *A Political Space* (2003, 78) reflected that

> when protesters talk to reporters, they are all values, justifications and pious outrage. This is as it should be; not everyone knows their reasons as well as they do and this is a way of reaching out. When they go back to camp, however, they talk little of those things ... In camp, there are strategic decisions to make, not to mention a thousand logistical details to be resolved.

Yet Braun paid curiously little attention to environmentalists' *strategies,* to the role of the media in framing events, or to environmentalists' dissatisfaction with media coverage. He appears to take media representations of the campaign for the campaign itself, or for the whole campaign. He also seems to imply that environmentalists had some control over the media and could have produced different framings of events in Clayoquot. At the same time his accounts of artefacts appear curiously abstracted from the everyday life of campaigning, and from the fact that there was no one campaign but rather a range of groups involved, with different locations and strategies. The specificities of the varied approaches taken by the different groups is not clear in his account, which does not distinguish between environmental groups, their locations (metropolitan or local), and the impact these locations had on the actions in which the groups engaged.

In writing "that few First Nations joined protestors on the blockades is a topic that has still not received the attention that it has deserved," Braun (2002, 8) seemed to imply that this was never discussed within the FOCS or at the camp, which, as Chaloupka recognized, was hardly the case. Braun's argument also relied on the implied assumption that the First Nations *should* have been on the blockades. But making them visible was not in the power of environmentalists, given that they had limited influence and little control over the media framing of the controversy. Nor was it their intention to do so. In a situation where everyday life in Tofino brought encounters in the street with loggers and First Nations, turning the media spotlight on the local would have further stressed already complex relations that the FOCS (and others) was trying to hold together. Catapulted into the media spotlight, the FOCS understood only too clearly the limits of such exposure – and its pressures. But it is also true that the FOCS recognized that justice, not greater visibility, was the appropriate response to the absence of First Nations. Justice would not necessarily be served through bringing "the media" to bear on First Nations communities in Clayoquot. Rather, it might be better realized by trying to reduce the stress of the media glare in the region – but also by spectacularly making visible histories of colonialism and by calling the Canadian state to account for its role in global deforestation. The ready and rather hyperbolic denouncements of the FOCS and of those arrested at Clayoquot as traitors demonstrates only too clearly how their actions were perceived by others. At the same time, Braun's arguments about the marginalization of First Nations were not entirely borne out in the aftermath of the campaign: First Nations were involved in the Interim Measures Agreement and its subsequent extensions, in the Central Regional Board, and in setting up Iisaak Forest Resources, a First Nations–led forest service company. Nonetheless, the limits of these developments also reveal the shortcomings of a politics of visibility or integration. These interventions hardly constitute appropriate or sufficient redress, and they also make clear that the First Nations cannot be understood as a homogeneous group.

Thus in pointing out that the turn to the global had rather other implications, I am also suggesting that this was less a puzzle than Magnusson and Shaw propose. I see their approach, reading Clayoquot through the global, as echoing the strategies of the FOCS, a direction which makes some sense when paying attention to the role of eco/feminism in the campaign. I suggest that it is not incidental that many people who joined the FOCS in the late 1980s and early 1990s were feminists. Although clearly not every FOCS member in the late 1980s explicitly identified as a feminist or eco/feminist, feminism was a shared concern for many, and it informed key decisions about how the camp and the campaign were organized. These feminists also came with a strong sense of contemporaneous feminist debates – an awareness of the limits of easy claims to global or even local sisterhood, an attention to the politics of location, situated knowledges, and the critiques of black and Third World feminists, and an understanding of their impact on white feminists. In particular I suggest they were well aware of the limits – that is, *dangers* – of a politics of visibility, or of giving voice. At the same time I would argue that eco/feminists insisted on continuing to articulate a politics of global connection at a time when many feminists were disengaging from such politics. During the 1980s and 1990s, even as white feminists were confronting the challenges of black and Third World activists and theorists, environmentalists were successfully questioning the relevance of national boundaries for containing environmental problems and were insisting that solutions to such problems also required international mediations. Similarly, they recognized the need for activism that transcended national boundaries. Though aware that they were implicated in histories of colonialism, many shared a profound sense of a feminist politics of the limits of home, which corresponds with what the late Val Plumwood (2008, 141) called an ethics of place. Unlike the reified sense of place of some environmental writing, Plumwood's ecofeminist ethic of place makes visible, not people in the Third World or indigenous peoples, but "north/south place *relationships,* where the north/south pole operates as a correlate of (various kinds of) privilege." This is the kind of work in which the FOCS has been engaged, work which began not with abstracted notions of wilderness or rainforest, but with Clayoquot, with attention to the specificities of the politics of location and situated knowledges.

Re-Vision as Eco/Feminist Genealogy

I have had many guides in providing an account of the eco/feminist politics of Clayoquot, and in seeking ways around the constraints of essentialism.

In particular, I mobilize a wonderfully generative tool – genealogy – to multiple ends, working genealogy against itself, as it properly demands. For many, "genealogy" will undoubtedly bring Foucault (1984) to mind, but my version of genealogy is more explicitly informed by feminist, queer, environmental, and postcolonial theorists and essayists (see also Tamboukou 2003). Genealogy is a much more productively promiscuous tool than Foucault allows, turning up in diverse locations and often undoing lines of supposed purity. It is particularly useful in a site where the influences of pioneering settlers/colonizers and earth mothers are supposed to abound. In weaving together its many sites and possibilities, I aim to develop the potential of genealogy as a generative method in telling stories of Clayoquot.

My formulation of genealogy draws on the work of postcolonial and anti-racist theorists such as M. Jacqui Alexander and Chandra Mohanty, Julia Emberley, Ann McClintock, Catherine Nash, Adele Perry, Mary Louise Pratt, Alice Walker, and Vron Ware. The scholarship of environmentalists such as Éric Darier, Paul Rutherford, Anna Tsing, and Terry Tempest Williams is also useful, as is that of Bruce Braun and Jocelyn Thorpe, who work specifically on Canada. Eco/feminists, feminists, lesbians, and queer theorists Judith Butler, Teresa de Lauretis, Sarah Franklin, Clare Hemmings, Alison Stone, Noël Sturgeon, and Maria Tamboukou have also influenced me, as have Susan Griffin, Donna Haraway, Adrienne Rich, Vandana Shiva, and Starhawk (though the latter do not explicitly use the term "genealogy"). Yet, creating a typology of genealogies would be inappropriate because genealogy is a much more tangled, knotted approach than this would imply; we would be better served to trace – that is, to construct – genealogy's own genealogy.

A key inspiration for me is the essay "When We Dead Awaken: Writing as Re-Vision (1971)," by the late Adrienne Rich (1979, 35). Clearly aware of the constitutive power of histories and the (re)writing of these histories, she stated,

> Re-vision – the act of looking back, of seeing with fresh eyes, of entering an old text from a new critical direction – is for women more than a chapter in cultural history: it is an act of survival. Until we can understand the assumptions in which we are drenched we cannot know ourselves. And this drive to self-knowledge, for women, is more than a search for identity: it is part of our refusal of the self-destructiveness of male-dominated society. A radical critique of literature, feminist in its impulse, would take the work first of all as a clue to how we live, how we have been living, how

> we have been led to imagine ourselves, how our language has trapped as well as liberated us, how the very act of naming has been till now a male prerogative, and how we can begin to see and name – and therefore live – afresh. A change in the concept of sexual identity is essential if we are not going to see the old political order reassert itself in every new revolution. We need to know the writing of the past, and know it differently than we have ever known it; not to pass on a tradition but to break its hold over us.

Notably, Rich's essay was first published in 1971, the same year that Foucault's "Nietzsche, Genealogy, History" appeared. Tracing genealogy's own genealogy through feminism, rather than exclusively through the patrilineage of Foucault and Nietzsche, is of course precisely genealogical. As Julia Emberley (2007, 239) noted of Foucault,

> naming his oppositional strategy a genealogy is, of course, somewhat ironic, especially when considered in relation to the filiative legacies drawn up in such histories of "mankind" as the Judeo-Christian Bible. The patrilineal accounts of father/son inheritances are seemingly endless and, in their exclusion of mothers, daughters, and sisters, function as a standard bearer of the autochthonous "search for origins."

In invoking Nietzsche, Foucault instigated his own genealogy, demonstrating the links between it and patrilineal knowledge practices. Thus, genealogy does not always undo the naturalization of knowledge practices. Emberley (ibid.) identified Foucault's own exclusions: "For to erect a patrilineal genealogy of pure descent is to assume that what has been excluded – the female reproductive body – is its natural, as in taken for granted, correlative, and, therefore, a priori excluded from the domain of culture and society." As I suggest later, eco/feminist matrilineages might properly be understood as genealogical – as a critique of patrilineage, as the construction of feminist knowledge practices, and as the intentional building of a matrilineal community of feminism/feminists that refuses kinship based solely on blood. A feminist genealogy reminds us that personal and political histories are entwined. Here Emberley's conjoining of patrilineal knowledge practices with recognition of the complicity of the patriarchal nuclear family in colonial and imperial projects is key.

Rich's method of re-visioning and her attention to "when we dead awaken" have a renewed relevance in the context of the rhetoric about the death of feminism and the related attention to writing histories and genealogies of the recent feminist past. I am also guided by Katie King's work, even if I cannot rise to the challenge of her careful, creative storytelling and capacity to (re)imagine both the world and writing differently. Her work on rethinking the terrain of cultural feminism is certainly useful for eco/feminists, given that both have been charged with essentialism (Alcoff 1988; Echols 1989; cf. Sturgeon 1997, who links cultural feminism and eco/feminism). In seeking to advocate for the term "cultural feminism," King (1994, 92) described her approach modestly: "I do not do this by answering Echols or other critics of cultural feminism and thus remaining in the epistemological paradigms she's constructed, borrowed, and exemplifies, allowing their hegemony, but rather by mapping an altogether different terrain," which she calls "the apparatus for the production of feminist culture." My account of eco/feminism is much more bound up with essentialism and related epistemological paradigms than I would wish, thus not quite undoing their hegemony. At the same time, I hold on to King's emphasis on the conversations and travels of feminist theory and the multiple sites where it might be enacted, in the hope of releasing essentialism's deadly grip on eco/feminism. To my practice of genealogy, I also bring King's focus on what she terms "pastpresents." As genealogy examines the history of the present, pastpresents pay attention to the persistence of the past in the present. These complicated entanglements of time are useful for telling stories of Clayoquot and of eco/feminism, not least because King (2010, emphasis in original) tied her pastpresents to Donna Haraway's work: "Postulating that pastpresents are a species of naturecultures, it offers many linked, or ... *knotted,* examples of how the past and the present continually converge, collapse and co-invent each other." Aware that "the past and present cannot be purified each from the other – they confront me with interruptions, obstacles, new/old forms of organization bridges, shifts in direction, spinning dynamics," King (ibid.) wanted to find ways to tell stories that do not rely on such purification. Her commitment to pastpresents, and to history as re-enactment, echoes Anna Tsing's commitment to friction and evokes Daphne Marlatt's *Ana Historic* (1990). As Marlatt explained about her novel,

> I like rubbing the edges of document and memory/fiction against one another. I like the friction that is produced between the stark reporting of document, the pseudo-factual language of journalism, and the more emotional, even poetic, language of memory. That's why I used such a hodgepodge of sources in *Ana Historic:* a little nineteenth-century and very local

> journalism that sounds like a gossip column, a 1906 school textbook, various historical accounts, some contemporary feminist theory, and a school teacher's diary from 1873 that was completely fictitious. (quoted in Kossew 2003, 56)

In this way, Marlatt reminds us of the stitched-together nature of history, of the craft involved in bringing different sources meaningfully together, and the necessary and productive f(r)ictions of re-visioning history in order to reimagine futures.

Eco/Feminist Genealogies: Denaturalizing Essentialisms

One of the challenges of articulating "nature" is the double meaning of the word: it refers apparently paradoxically to what is not human but can also denote that which is precisely human, as in human nature – and women's nature. Moreover, there is often a slippage between the two meanings. Thus, arguments about women's nature have commonly been used to restrict their lives and have led some feminists to insist on distance from nature, as opposed to a re-examination of both women and nature. Consequently, eco/feminist insistence on thinking about women and nature, and feminism and nature has not always been seen as being truly feminist. Rather, it has been thought to undermine feminism because its attention to nature is perceived as essentialist. Teresa de Lauretis (1989, 3) offers a useful account of feminist theory that helps to elucidate some of eco/feminism's challenges: "Feminist theory is all about an essential difference, an irreducible difference, though not a difference between man and woman, nor a difference inherent in 'woman's nature' (in woman as nature) but a difference in the feminist conception of woman, women and the world." By understanding eco/feminism as a reworking of feminism, where the world that de Lauretis mentions is taken up quite literally as planet earth, we can see that eco/feminism properly understands itself as both "of feminism" and a critique of it. Thus, one of the arguments I make throughout is that issues in eco/feminism are often feminist dilemmas writ large, intensified because of the doubling of the "natures" at stake; and it is the attention to the multiplicity of nature that renders eco/feminism particularly fruitful in examining developments in feminism.

In her account of eco/feminism, Jenneth Parker (2001) suggests that naturalism, not essentialism, is the issue that requires scrutiny and that concentrating on this would achieve what feminists and eco/feminists desire. What Parker has pointed to is that nature is taken up in particular ways at certain times, as is the case for essentialism and its variants such as biological determinism and universalism. As a result, essentialism rather than naturalization becomes the focus. This point is vital in my use of genealogy as a methodological approach which works at denaturalization. So, rather than going "essentialist hunting," as Stengers (2008) usefully puts it, alluding to the witch hunts, I trace processes and moments of denaturalization. In doing so, I invoke an eco/feminist genealogical practice to explore how we can undermine the critique of essentialism and how we can learn to ask the question of nature anew.

Turning to the specific, to Clayoquot, makes recourse to abstractions difficult. The particular enables the refusal of essentialisms. Ultimately, we do little justice to eco/feminism by thinking in terms of essentialisms or of closeness to nature. There are, and have been, other ways of thinking about eco/feminism, one of which is as an anti-dualistic politics of interconnection, though curiously, this understanding is often forgotten. Barbara Gates (1998, 20), to cite just one example here, noted simply that "inherent in ecofeminism is a belief in the interconnection of all living things. Since all life is nature, no part of it can be closer than another to 'nature.'" Hence, it makes little sense to talk of eco/feminism in terms of women's closeness to nature. That is not to say that we cannot acknowledge the circulation of discourses of women as being closer to nature. Rather, this does not produce a useful account of eco/feminism. Nonetheless, we might take the strength and persistence of such accounts as something important to explore.

Throughout this book, I use genealogy to undo essentialisms and to produce alternative stories of eco/feminism. I begin to trace the implications of a genealogical approach in Chapter 2, taking up Noël Sturgeon's effort to unpick typologies of eco/feminism. I also extend Sturgeon's genealogy, revisiting accounts of eco/feminism that proliferated *before* the reductive turn to typologies. In particular, I highlight the rather different places occupied in the eco/feminist literature by the Indian eco/feminist Vandana Shiva and the Chipko Movement, and the reported naming of eco/feminism by French feminist Françoise d'Eaubonne. I juxtapose these different trans/national stories with a reflection on Shiva's appearance at a conference on Vancouver Island and histories of H.R. MacMillan's travels from British Columbia to India to find a market for timber a hundred years ago. In tracing the increasing marginalization of activism, I seek to address *how* and *why* theories of essentialism came to dominate questions about women and nature.

Questions such as whether women are closer to nature than men are explicitly rejected in Chapter 3. Instead of asking what should our relationship with (an external) nature be, I ask what is our relationship with ourselves? Turning to oral history interviews with Clayoquot activists as a site of eco/feminist theorizing, I use the possibilities of narrative for disrupting essentialisms and lay the groundwork for my argument for understanding these stories as "unnatural histories," as more-than-human genealogies of the self, which insist on kinship with the earth. Through the intentional construction of eco/feminist genealogies, I recuperate the practice of tracing eco/feminist foremothers, which is dismissed by others as a vestige of matrilineal and reproductive thought. In this way, I also take up genealogy's "other" life as family history, linking such histories with genealogical knowledge practices. This approach is in keeping with my interest in complicating genealogy as a critical practice of knowledge production and my refusal to separate out the making of family, and family histories, from "global"/universal metahistory – and indeed from natural, and naturalized, history.

Chapter 4 turns to a genealogical approach to Clayoquot, exploring the emergence of the campaign in the context of local and regional histories of Vancouver Island and British Columbia, including fictional accounts. I argue that one of the accomplishments of the Friends of Clayoquot Sound (FOCS) is its challenge to patrilineal histories of the forest industry and forest science, and colonial histories of Clayoquot, and its production of alternative histories, which point to possible different futures. In doing so, I point to the *cultural*, as well as environmental, work of the FOCS. The FOCS highlights culture as an important domain of politics, that not all politics is fought on blockades, or through ballot boxes, but that some of the most important political work is the transformation of the Canadian imaginary.

Chapter 5 turns to the summer of 1993, to the Clayoquot Peace Camp and the blockades in which over eight hundred people were arrested. This chapter addresses one of the paradigmatic examples of essentialism, that of non-violence, through exploring the eco/feminist politics and the gendered politics of everyday life at the camp. The camp was a moment when eco/feminist politics came sharply into focus in the campaign, where eco/feminism emerged, was contested and perhaps temporarily disappeared. Although the camp was said to be based on "eco/feminist principles," links between feminism, non-violence, and consensus were not always transparent to participants. The chapter addresses how meanings of eco/feminism were articulated and defined through attention to process and practice, rather than as an identity politics. Contestations over the practice of eco/feminist politics are addressed by narrating the events of the summer through the prism of conflicts over gender and eco/feminism, by raising questions about constructions of masculinity in the environmental movement, gendered conflict, tensions around the practice of consensus and non-violence, and the meanings of safety and security at an eco/feminist camp.

An engagement with the lives of women activists at the camp provides the basis of Chapters 6 to 8. Collectively, these chapters work to undermine many of the assumptions about eco/feminism – and especially about maternalism – and they point to the importance of empirical research in this context. In the Preface, I have already confessed that I was not initially that interested in trees. Now I add that I was not initially that interested in motherhood either. To be clear, at the same time I did not accept the story that eco/feminism was all about maternalism, and indeed the research was to a certain extent motivated by the assumption that something more/else was going on in women's environmental activism that could not be reduced to maternalism. Yet, though I did not think motherhood was enough, at the same time I did not think it was irrelevant, and eventually I realized that writing about women's efforts to renegotiate motherhood was indeed a key part of many women's stories, including those who did not have children.

Thus Chapter 6 takes up the thorny question of that supposedly paradigmatic manifestation of essentialism, maternalism. Here, I provide an altogether other account of motherhood in eco/feminist politics to any available thus far. In doing so, I demonstrate the value of an approach that does not persist in seeking essentialism, but rather pursues ways of listening attentively to women's stories of their lives. Rather than an essentialist over-determination of motherhood, I found a profoundly eco/feminist effort to rework meanings of motherhood, family, and home, which calls into question any simple reduction of eco/feminism to extending maternal concern about children to the planet, and also acts as a reminder of the importance of "a politics of location." That these stories are told by white women in Clayoquot Sound acts as a salutary reminder to me of the importance in reconfiguring meanings of family and home, and how this has always been key for feminists. As Emberley, Perry, and others remind us, we must not forget the role the reproduction of the white nuclear family has played in the making of empire, and paying attention to sites where the heterosexual patriarchal nuclear family is undone, and how any undoing of empire and colonialism also requires the undoing of the bourgeois colonial family, remains a central feminist project.

In Chapter 7, I go on to juxtapose the overdetermination of motherhood in accounts of eco/feminism with a near complete lack of attention to childhood. This absence contrasts markedly with the reification and romanticization of childhood in the writings of deep ecologists. The chapter argues that, in order to reconceive motherhood, childhood also needs to be retheorized. In my recounting of women's recollections of childhood natures, I disrupt the tendencies of some deep ecologists to produce romantic origin stories of happy childhoods in nature as an origin story of adult activism. For many of the women with whom I spoke, no nostalgic return to childhood was possible. Adult activism offers a complicated working through of gendered childhood natures. Childhood, in particular for girls, is often a site of conflict, where nature is used to mark the boundaries between child and adult, girl and woman, and where nature is invoked by adults in gendering children. Girls learn to be girls through, for example, differential access to nature and the "outdoors," compared with boys. The chapter pays particular attention to the boundaries between home, "nature," and the social world, and the role of families in negotiating these boundaries.

Chapter 8 draws together the radical potential for eco/feminist activist narratives to reconfigure the meanings of motherhood, childhood, family, home, genealogy, and kinship. I suggest that in telling their life stories, women produce genealogies of a more-than-human self, breaching the boundaries of human and nature/planet. These narratives involve a fundamental re-creation of family and home. Genealogy is conventionally understood as the construction of a family tree, a way of mapping kinship, of blood relations and marriage ties. Yet, these women's stories can be understood as efforts to reconstruct familiar and familial genealogies. The stories provide unnatural histories of decidedly unfamiliar "family trees" (Williams 1991) and invoke relatives, friends, and mentors as part of a newly reconstructed family. For many of these women, working through relationships with home has also meant working through the meanings of being "at home" in Canada, and specifically in Clayoquot Sound. If Clayoquot is home, it is one that has had to be made and struggled for. Making home, for these women involves "home-work." The work of living in Clayoquot involved trying to resist being a settler, trying not to be implicated in the destruction of nature, but taking seriously the question of how to make a home in nature. The women of Clayoquot Sound were engaged in the work of the co-construction of nature, which sometimes involved blockading logging roads and going to jail. For many, the campaign, the blockades, and subsequent court proceedings became a site of engagement with and contestation over national identity and belonging, over the meanings of being a Canadian citizen. Thus, the reworking of home demanded that women forge an ethical and ecological citizenship, a reconstructed Canadian identity.

The themes of motherhood and sexuality, childhood, meanings of family and home, including Canada as home and as national identity, which occupy Chapters 6-8 focus on aspects of women's lives that are frequently naturalized. Foucault (1984, 76) wrote that genealogy

> must record the singularity of events outside of any monstrous finality; it must seek them in the most unpromising of places, in what we tend to feel is without history – in sentiments, love, conscience, instincts; it must be sensitive to their recurrence, not in order to trace the gradual curve of their evolution, but to isolate the different scenes where they engaged in different roles.

Hence my return to what might appear to be the most unpromising of places – an eco/feminist peace camp, motherhood, and childhood, all utterly overdetermined parts of women's lives and troubling sites for eco/feminists. Together, the chapters on motherhood, childhood, families and home allow me to understand women's oral histories as a genealogical project, one where retelling family history and the refusal of a search for origins opens into a genealogy of a more-than-human self.

Following this journey of reconfiguring eco/feminism through Clayoquot, I return in the final chapter to reconsider some of the animating concerns of this book, including the contrast between the vitality of eco/feminist activism and the persistence of narratives of the end of feminism. I note that a vocal insistence on eco/feminism's anti-essentialism, or strategic essentialism, has not secured eco/feminists a place at the feminist table. My thinking on the "nature" of feminist history is provoked by Clare Hemmings's work on "telling feminist stories." Hemmings identified the dominant narratives of feminist history as progress narratives, or stories of decline or loss. Key to her analysis is attention to periodizations of feminist histories – how decades come to stand for particular issues. The nineties were a site of debate over difference and the seventies the proper time of essentialism. Yet eco/feminism as an explicit movement emerged in the late 1980s and 1990s; and with its attention to women and nature, and assumed essentialism, eco/feminism emerges as out of step with dominant periodizations of feminism – as essentialist at the wrong time, at a time when all feminists should have known better than to reproduce essentialisms. This analysis begins to reveal why no amount of anti-essentialism, or even strategic essentialism, will lead to a place at the feminist table. Although the emergence of new feminisms, such as new feminist materialisms, appears to offer a (re)consideration of matter, I point to how they too appear complicit with Hemmings's narratives of linear displacement. Against such feminist histories, I call for knotted histories where contradictions might remain coeval, at the same time, where eco/feminism might yet be understood, not as a return to the essentialist seventies, but rather as a site of ongoing reflexivity over "women" and "nature."

Genealogy, of course, also requires proper attention to limits. This is a more-than-human genealogy, but it is not a multi-species ethnography (Kirksey and Helmreich 2010). I do not foreground the other species of Clayoquot or the other species in the lives of those I interviewed – nonetheless, I do try and hold other species in sight, in the frame, so animals are mentioned in the stories I tell. At the same time, neither do I want to leave the human – that is, the humanist, or human exceptionalism – unchallenged. In fact, one concern is that the broader turn to animal studies can leave human/(non-human) animal dualisms entirely intact. Here I am distinguishing this more recent turn to animal studies from the long eco/feminist commitment to animals (Adams 1991; Gaard 1993b, 2002; Kheel 2008), which has always had a more relational focus. I am interested in producing an account of the more-than-human where, contrary to general usage, the more-than-human does not collapse into the non-human, but rather draws our attention to how we might think the human otherwise, how the human might be reconfigured through the process of activism, dwelling, and telling stories. I am interested in how paying attention to stories and genealogies can produce entirely unnatural histories where invocations of Mother Nature, rather than being an utterly overdetermined figure, might be read as an effort to claim kinship with the earth.

This book is hardly the whole story of Clayoquot or even of the interviews I heard. The book draws on interviews with women who were involved in the campaign in various ways, many of whom were at the camp and were arrested. It also draws on a small number of interviews with men involved, as well as a group interview. I did not interview anyone who explicitly identified as First Nations; nor did I interview anyone in the logging industry. Though I draw extensively on extended quotes, as everyone who attempts oral history discovers, there is no way to do justice to the material of people's lives. Any account is thus inevitably a kind of failure. In the meantime, this book is an attempt to tell a very particular story of Clayoquot, at a particular point in time – as Antoinette Burton (1992, 26) argued, "The history we are writing and the theory we are reading are themselves the products not just of our cultural milieux, but of the historical moments we are living in as well." There are still many stories to be told about Clayoquot, and I hope that I, and others, can find ways to continue to do this.

2

Eco/Feminist Genealogies

Essentialism, Universalism, and Telling (Trans)National Histories

Ecofeminism makes such big promises!

– STEPHANIE LAHAR, "ECOFEMINIST THEORY AND GRASSROOTS POLITICS" (1991)

I first encountered the word ecofeminism in 1987 when I was a master's student doing research for a term paper in a course in feminist sociology. Freshly arrived in the big city of Toronto, which I then perceived to be completely devoid of nature (I grew up in Victoria on Vancouver Island, where nature may not have been more plentiful but was certainly bigger), I was absolutely thrilled to discover a word that already existed to represent my deepest personal and political desire, the inclusion of an environmentalist perspective in feminist theory. I craved a language that would describe my growing sense that nature must be an important consideration in any feminist political vision; I remember devouring the first ecofeminist text I encountered – *Green Paradise Lost,* I think it was – at the expense of the readings on public policy on which I was supposed to be focusing.

> But the exhilaration I felt as a new convert was over quite soon, and I have never felt so strongly that I belonged in ecofeminism, despite my increasing commitment to feminist ecological politics and theory.
>
> – CATRIONA SANDILANDS, *THE GOOD-NATURED FEMINIST* (1999)[1]

> Although I've been attracted to thinking at the intersections of feminism and environmentalism for years, I hesitate to call myself an ecofeminist. Indeed, I prefer to think of my work as ecological feminism, in an effort to keep the emphasis on feminism, and also to distance my approach somewhat from other work done by self-titled ecofeminists. Though I share motivations with the authors of such work, I am sufficiently critical to be uncomfortable with the label. Accordingly, in these pages, "ecofeminism" is an umbrella term referring to forthright attempts to link some versions of feminism and environmentalism, and "ecological feminism" refers to a specific subset of ecofeminist approaches I wish to articulate and endorse here.
>
> – CHRIS CUOMO, *FEMINISM AND ECOLOGICAL COMMUNITIES* (1998)

> Dealing with my own objections to the essentialism of some ecofeminist arguments, and the effects on my work of a widespread assumption by academic feminist peers that such essentialism permanently and thoroughly tarnish ecofeminism as a political position, I have struggled with the question of whether I would want to identify myself and my work as ecofeminist.
>
> – NOËL STURGEON, *ECOFEMINIST NATURES* (1997)

Confessions of Reluctant Eco/Feminists: Essential(ist) Concerns

Eco/feminism, once described with excitement as offering such promise (see also Warren 1990), had by the mid- to late 1990s become for many in academia a somewhat "reluctant confession" (Cuomo 1998, 5) or even an identification that some wished to disown entirely.[2] This shift from exhilaration to disavowal speaks of an intense and fraught period in eco/feminist history. Only a few years separate Lahar's excited comment – "Ecofeminism makes such big promises!" – from the struggles of Sandilands, Cuomo, and Sturgeon with the term "ecofeminism." For them, as for many others, reservations about essentialism came to circumscribe possible identifications with eco/feminism. Or, more precisely, their concern was that they might be perceived as essentialist by other feminists. To be clear, Sandilands, Cuomo, and Sturgeon did not straightforwardly accept that eco/feminism is essentialist. Indeed, all robustly provided accounts of eco/feminism as that which cannot be reduced to essentialism. Nonetheless, the weight of essentialism marks the work of many eco/feminists. At the same time, whatever label they might choose, all resolutely continue to work at the interfaces of "women" and "nature," feminism and environmentalism. This chapter explores the palpable disappointment of the 1990s onward and why such hope turned to such disillusionment. In particular, I focus on how essentialism and critiques of universalism figured in this shift in accounts of eco/feminism in the 1990s. In doing so, I also highlight accounts of eco/feminism that have not been so overdetermined.

My "confession," or as I would prefer, *affirmation* (following V. Bell 1999), is that I am, *still,* a passionate eco/feminist. I have no doubt that my ongoing enthusiasm for eco/feminism is in no small part due to my experiences at Clayoquot. It is worth noting that the Clayoquot camp and blockades occurred in 1993, in the time between the initial excitement regarding eco/feminism and the subsequent disenchantment. Nonetheless, I write some time after Sandilands, Cuomo, and Sturgeon, or more accurately, I write before, during, and now finally more than ten years after their "confessions." Now is a different moment in which to make an affirmation of eco/feminism, even if accusations of essentialism might still be forthcoming. Although many years on, it still feels as if concerns about essentialism continue to haunt eco/feminism, and clearly this book is marked by the intractable nature of the debate. Sturgeon (1997) reported encountering audiences that anticipated (even demanded) either essentialism or anti-essentialism, and she reflected on her attempts to confound both expectations. I have sometimes experienced something slightly different – a sense of puzzlement about why I still feel mired in a debate that others probably perceive as anachronistic. This perception suggests a failure to appreciate the damage caused by the critique of essentialism. The disavowal of eco/feminism, by feminists and by eco/feminists themselves, might be understood as a consequence of this critique. Eco/feminism thus offers a particularly useful site for an examination of the devastating impact of the accusation of essentialism, as well as of an insistent anti-essentialism, on feminist theorizing and activism. The quotes that open this chapter demonstrate that talking about eco/feminism (in an academic context at least) had become difficult by the late 1990s unless one also addressed the inevitable question/accusation of essentialism and positioned oneself – critically – in relation to it. In the context of the hyper-politicized "nature" of essentialism, eco/feminists have been frustrated that their research and politics have been dismissed as essentialist and that others have doubted their feminist credentials. For instance, in *Ecofeminist Natures,* Noël Sturgeon (1997, 167) stated, "Some ecofeminists have argued that there is an 'establishment feminist backlash' against ecofeminism, resulting in a lack of ecofeminist writing in prominent feminist journals such as SIGNS, as well as the invisibility of ecofeminist theory in important academic feminist conferences and anthologies." Offering examples from her own career, Sturgeon (ibid., 6, 167-68) recalled that a feminist mentor advised her to delete "eco/feminism" from the title of one of her papers and that another suggested that her job applications should not mention her editorship of the *Ecofeminist Newsletter.* Such anecdotal accounts of feminist distancing of eco/feminism, together with eco/feminists' sense of rejection and unjustified exclusion from the feminist sisterhood, begin to suggest just how much has been at stake and gestures to the experiences that underlie the opening quotes of this chapter.

In constructing a story of eco/feminism and feminism, I am developing Noël Sturgeon's genealogical approach. Necessarily, in the process I generate my own – interested – genealogy of eco/feminism. Indeed, this is one of the appealing features of genealogy, the inevitability of being implicated in the story. Sturgeon's work provides an account of dominant practices of describing eco/feminism through typologies at a specific juncture – the early and mid-1990s. Throughout, I circle around Sturgeon's generative work, trying to expand the realms of possibility for eco/feminism beyond essentialism and anti-essentialism, while also suggesting the limits of strategic essentialism. One of my aims is to provide an account of eco/feminism before essentialism became such a determining presence, in order to gesture toward alternative and more productive understandings, and to explore what has been at stake in eco/feminists' efforts to divest themselves of critiques of essentialism and universalism. In particular, I turn to eco/feminist origin stories and juxtapose the citation of Françoise d'Eaubonne with the iconic place of the Chipko Movement and the work of Vandana Shiva. Through exploring a moment when Chipko was invoked in the

Clayoquot campaign, I offer an account of eco/feminism that opens up the possibilities of transnational theory and activism, rather than collapsing into national narratives or even silence about other parts of the world.

Disowning Eco/Feminism: Essentialism, Anti-Essentialism, and Typologizing

Sturgeon (1997, 4) described her account of eco/feminism as a "genealogy rather than a history." Her take-up of the possibilities afforded by genealogy offers a more nuanced discussion of the implications of different descriptions of eco/feminism. In doing so, she has made an invaluable contribution to articulating eco/feminism's project. As Sturgeon (ibid., 3-4) explained,

> My version of the history of the origins and development of ecofeminism is thus not so much a coherent narrative of the even, dependable growth of an independent political position as it is several snapshots of scattered, uneven, and in many ways disconnected beginnings, retreats, dormancies, and proliferations imbedded within several different political locations. This is a genealogy rather than a history, and as a result, I am not following one unitary subject (ecofeminism) through different historical moments. Instead I am articulating relationships, legacies, simultaneous births of related entities, discontinuities, renamings, mutations, throwbacks.

In constructing her own genealogy, Sturgeon traced the emergence of eco/feminism through its appearance across a range of sites, academic, activist, organisations, and writings of all kinds. Her careful attention to ways of describing and defining eco/feminism is instructive. Sturgeon's method offers a counter to the typologies that she documented in the final chapter of *Ecofeminist Natures*. Here she thoroughly scrutinized the practice of creating eco/feminist typologies – radical, cultural, and socialist – and of generating new terms such as "feminist ecological politics," "ecological feminism," and "environmental feminism." She clearly identified typologizing as an exercise in naming and exclusion, as enabling the disavowal of certain aspects of eco/feminism by carving it up, disowning "essentialist" varieties and renaming "good" (non-essentialist) ones. In this process, she revealed what was at stake in the search for a new descriptor for ecofeminism that was purified of any taint of essentialism.

Sturgeon identified two common effects of typologizing in eco/feminism. First, she pointed out that feminisms of colour are frequently ignored. Second, citing Chela Sandoval (1991), she noted that superior and inferior types of feminism are implicitly constructed, making coalitions between activist and academic feminisms difficult. Sturgeon (1997, 16) argued that

> the critique of essentialisms of various kinds has been a prominent tool in creating various types of typologies of feminisms, usually to support an "agonistic narrative structure" in which certain feminist theories (usually socialist feminism or poststructuralist feminism) come out to be the winners in the contest for the most politically useful feminist theory. That these winners in the anti-essentialist competition have also been the feminist theories most embedded in the academic contexts is suggestive.

Sturgeon's analysis offers ways of understanding much about eco/feminism's current state. I want to take her genealogical approach further, to see what might be learned from applying it to other ways of describing eco/feminism.

Diverse Eco/Feminisms?

Typologizing has not been the only approach to defining eco/feminism. During the late 1980s and early 1990s, the emerging and overlapping academic and activist literatures commonly described eco/feminism by stressing its diversity. In fact some suggested that, precisely because of its pluralism, characterizing the movement in a simple way was impossible. For instance, Elizabeth Carlassare (1994, 220-21) wrote,

> Ecofeminism does not lend itself to easy generalization. It consists of a diversity of positions, and this is reflected in the diversity of voices and modes of expression represented in ecofeminist anthologies. The ecofeminist anthologies, *Reclaim the Earth,* and *Reweaving the World,* and the issues of *Heresies* and *Hypatia* on feminism and ecology include the work of different women from different countries and social situations, and their work does not adhere to a single form or outlook. Poems, art, photographs, fiction, prose, as well as theoretical/philosophical/"academic" works are included. Ecofeminism's diversity is also reflected by its circulation in a variety of arenas, such as academia, grass-roots movements, conferences, books, journals, and art.

This range – of positions, of voices, of forms, of locations – was viewed positively, as suggesting that eco/feminism was not dogmatic and could embrace difference, an important value in feminism during the early 1990s. With such an emphasis on diversity, anthologies and special issues of journals, which allowed for a variety of voices and forms, were the preferred publication formats at the time. Even so, their editors sometimes felt the need to point out their limitations, as in Judith Plant's introduction to the collection *Healing the Wounds: The Promise of Ecofeminism* (1989b, 4): "This anthology in no way fully represents the wide spectrum of thought that is ecofeminism."

However, some commentators suggested that this emphasis on diversity might create confusion. Stephanie Lahar (1991, 28) observed that "the newness of the movement, the breadth of issues it encompasses, and the diversity of people thinking and writing about ecofeminism have resulted in considerable confusion about what ecofeminism actually is, who ecofeminists are, and what they have to say." Thus, though diversity was prized, making sense of eco/feminism's sheer variety also constituted something of a challenge, and one feature came to be foregrounded – the emphasis on diverse locations all over the world. Even in the quotes above, it is possible to trace the privileging of the international. Carlassare (1994, 220) noted the inclusion in anthologies of "the work of different women from different countries and social situations," and Plant (1989b, 5) invoked "a truly international movement."

The importance of the international is particularly clear in the anthologies *Healing the Wounds* (Plant 1989a), *Reweaving the World: The Emergence of Ecofeminism* (Diamond and Orenstein 1990b), and *Reclaim the Earth: Women Speak Out for Life on Earth* (Caldecott and Leland 1983b). All three stressed the importance of global activism. As Caldecott and Leland (1983a, 5) explained, "In many countries all over the world, women are taking an increasingly prominent role in political struggles: in the peace, anti-nuclear, health and ecology movements." Diamond and Orenstein (1990a, vii, emphasis added) stated,

> This volume, with its chorus of voices reflecting the variety of concerns flowing into ecofeminism, challenges the boundaries dividing such genres as the scholarly paper and the impassioned poetic essay. In so doing, it acknowledges poetic vision as a form of knowledge and as one of the important steps in the process of *global transformation.*

> Ecofeminist philosophy and creation have brought about changes in consciousness, political action, and spiritual practice. This collection is composed of writings by poets, novelists, scholars, scientists, ecological activists, and spiritual teachers, many of which were first presented at the conference, "Ecofeminist Perspectives: Culture, Nature and Theory," held at the University of Southern California in March 1987 and sponsored by the California Council for the Humanities. We also invited others who did not attend the original conference to contribute to the volume, thus making a more comprehensive statement about *the global emergence of ecofeminism.*
>
> We hope to dispel the notion that poetry and politics, spirituality and activism, scholarship and vision are to remain forever divided, either from each other, or within the same person. The writers and activists represented here share *a multicultural and diversified global vision of healing for life on Earth.* Their voices have been inspiring to us, and our fervent wish is that this volume will be a catalyst for further change.

In her foreword to *Healing the Wounds,* Petra Kelly (1989, ix-x) wrote,

> This is a book about global ecological sisterhood! ... This is not a time for complacency. It is a time for continuing to link arms as sisters – like the women in the Chipko movement in India; like the women at Greenham Common, in England, who are not giving up the struggle against militarization; like the women of the Western Shoshone Indian Nation in Nevada who opposed nuclear testing by encircling the test grounds; like the women in the Pacific struggling for a nuclear-free future to prevent babies being damaged through French atomic bomb tests; like the women in the Krim Region of the Soviet Union demonstrating courageously against a new nuclear power plant.

Commonly, as Kelly demonstrates, eco/feminism was made manifest through listing places or organizations where eco/feminist activism was understood to be emerging. Thus, eco/feminists did not espouse just any international – they lauded women's grassroots *activism* around the world. Introducing eco/feminism by providing a list of activisms was common in the late 1980s and early 1990s (see, for example, Mellor 1992, 50; Merchant 1992, 184; Baker 1993, 2; and Gaard and Gruen 1993, 1). In addition to the examples cited by Kelly, these included the Kenyan Greenbelt Movement, the Women's

Pentagon Actions, the Narmada Valley dams campaign, the Love Canal Homeowners drive against toxic waste, the Chernobyl and Bhopal campaigns, Japanese women's consumer movements, the Women's Environmental Network, and many more. The diversity of locations in which the activism took place was as significant as the activism itself. Eco/feminists liked to say that the movement emerged at the same time in many parts of the world. Salleh (1997, 17) wrote of "the word 'ecofeminism' turning up spontaneously across several continents during the 1970s."

Yet perhaps the list eventually became too difficult. Perhaps, following Lahar's observation that the diversity of eco/feminism seemed to produce confusion, holding on to diversity became too challenging, with the result that one instance – the Chipko Movement – emerged as iconic. Chipko was a movement of Indian women in the Garwhal Himalayas who hugged trees to protect them from commercial logging. It was brought to prominence by scholar and activist Vandana Shiva, in her book *Staying Alive: Women, Ecology and Development* (1989), and also through her numerous presentations at international policy and activist conferences. Chipko has long been contentious in eco/feminism as a particularly problematic manifestation of essentialism, that of cultural appropriation and of universalizing women-nature relationships (see Moore 2008, 2011). The turn to typologizing, the concomitant reification of Chipko, and the reduction of eco/feminism to the Chipko Movement might then be understood as responses to the challenges posed by its diversity. Yet all these accounts – of diversity, of simultaneous emergence around the world, and of Chipko as exemplary – exist in tension with another account, one that is often visible only in the footnotes of eco/feminist texts. In the late 1990s, when Sturgeon (1997, 167-96), Cuomo (1998), and Gaard (1998, 12-15) echoed each other in asking "What's in a name?" they were referring to the various names that were being created to describe typologies of feminism and environmentalism. The naming that received less attention was that of French feminist Françoise d'Eaubonne as originator of the term "ecofeminism."

Françoise d'Eaubonne and the Origins of the Term "Ecofeminism"

The first use of the term "ecofeminism" (as *"ecoféminisme"*) is commonly attributed to Françoise d'Eaubonne, in her book *Le féminisme ou la mort* (1974). D'Eaubonne has been widely cited in the eco/feminist literature from the mid-1980s onward. It is as if this fact could provide some form of anchorage, when confronted with the multitudinous protests, books, courses, workshops, conferences, poems, and rituals that have come to constitute eco/feminism. The ritual and obedient citation of d'Eaubonne's act of naming testifies to the power of this moment for academics in particular and to the difficulties that they appear to have had in understanding and interpreting its significance. Yet, although d'Eaubonne's act of naming has been accorded some prominence in literature on eco/feminism, it has often been minimized at the same time, as a brief passing reference, or relegated to a footnote. It is as if d'Eaubonne might be credited with sole authorship, or ownership, of the new phenomenon, which would implicitly threaten both the authorship rights of others and the grassroots, internationalist, and activist basis of eco/feminism itself. Indeed, one of the few extended accounts of d'Eaubonne's work, by Barbara Gates (1998, 15-16), explicitly and unusually clarifies this point: "I would like to reemphasize the often repeated fact that ecofeminism involves activism as well as ideology and that both of these aspects arose simultaneously worldwide ... This essay is not, then, a 'culturalist' attempt to privilege European written discourse or theory over political action, or other legitimate manifestations of ecofeminism."

Yet, alongside the persistent citation of d'Eaubonne, the significance of her act of naming has also been disputed in a number of ways. Some, such as Sturgeon (1997, 202n8) and Gaard (1998, 13), have sought for the first reference to d'Eaubonne as originator of the term, as if in an attempt to apportion responsibility for what has been seen as a problematic version of eco/feminist history. They both identified Karen Warren's 1988 paper "Towards an Ecofeminist Ethic" as the source for the claim, and Gaard (1998, 13) stated that "from that point forward, many writers cited this attribution without verification." This apparently unthinking recitation supports the idea that d'Eaubonne's work had no substantive impact.

Given that *Le féminisme ou la mort* was not translated into English until much later, some have argued that its influence on the development of eco/feminism in the United States would have been negligible (Sturgeon 1997, 202n8; Heller 1999, 65; Sandilands 1999, 6). The relief at finding this resolution to the dilemma of authorship was almost palpable in one overview of eco/feminism:

> Françoise d'Eaubonne's *Le Féminisme ou la Mort?* (1974) is often credited for the first usage of the word "eco-feminism" but as Ariel Salleh has pointed out in *Hypatia,* Vol. 6, No. 1 (Spring 1991), pp. 206-14, esp. 206,

> the text was not translated into English until 15 years later, so it would have had little effect on those articles and texts published before 1989. The spontaneous appearance of the word or the concept of "eco-feminism" across several continents would indicate that it arose out of a real, internationally observable phenomenon rather than the influence of a single writer. (Gaard and Gruen 1993, 1)

The chronology of translation often plays an important role in these accounts. Although a short extract from the book was published in English in 1980 as "Feminism or Death" (d'Eaubonne 1980), it did not contain the word "ecofeminism," a fact that simply reinforced arguments about d'Eaubonne's lack of influence on early eco/feminism in the United States. Heller (1999, 65n20) cited 1994 as the date when d'Eaubonne's work became available to English-speaking audiences; in that year, a chapter from *Le féminisme ou la mort*, which did include the word "ecoféminisme," was finally translated and anthologized in a collection edited by Carolyn Merchant (d'Eaubonne 1994).

However, relying on matters of translation to determine d'Eaubonne's influence assumes that no one outside of France is capable of reading French and that little cross-cultural exchange occurs. Gaard (1998, 13) noted, slightly incredulously, another route for d'Eaubonne's work, acknowledging that "others, such as Carol Adams, claim that they had learned the term 'ecofeminism' from Mary Daly, whose work *Gyn/Ecology* had used it in 1978 and whose classes included a study of Françoise d'Eaubonne's text in the original language." In *The Sexual Politics of Meat*, Carol Adams (1991, 89) pointed out that d'Eaubonne's text was introduced "to scores of feminists who took Mary Daly's feminist ethics class at Boston College" just after it was published in 1974. Furthermore, Daly (1978, 9) explicitly referred to d'Eaubonne in discussing the naming of *Gyn/Ecology*, which was first published in 1975. Adams (1991, 89, 218n12) attributed the appearance of her own book, as well as Collard and Contrucci's *Rape of the Wild* (1988), to Daly's influence, Collard being a close friend of Daly. She also pointed to the broader influence of *Le féminisme ou la mort* and of eco/feminist ideas linking feminism, animals, and nature on discussions in the Cambridge-Boston women's community. The books by Daly (1978), Collard and Contrucci (1988), and Adams (1991) are among the early texts of eco/feminism, and Daly's *Gyn/Ecology* is also a central text of early second-wave feminism. This suggests that d'Eaubonne's work did have both a significant audience in the English-speaking world and an appreciable impact on the literature of eco/feminism, despite the continuing lack of an English translation. However, this hardly equates with attributing eco/feminism per se to d'Eaubonne alone and does not necessarily challenge claims for eco/feminism as a grassroots phenomenon that emerged in various parts of the world. It is possibly more an indication of the significance of Mary Daly in the genealogy of eco/feminism.

Furthermore, the attention to translation glosses over what was Salleh's main concern. Writing of the spontaneous rise of ecofeminism across the world, she made the point that "for politico-economic reasons then, ecofeminists working from more visible niches in the dominant English-speaking culture have tended to get their views broadcast first – even feminism is touched by its imperialist context." Salleh's (1991, 206) note that "the lonely appearance in Paris of Françoise d'Eaubonne's *Le féminisme ou la mort* (1974) is an exception whose lack of an English translation some fifteen years later, more or less proves the rule" was intended to highlight the dominance of the English-speaking world and the rarity of eco/feminist texts in a language other than English (see also Gates 1998).

Sandilands (1999, 6) astutely pointed to the need to recognize that "d'Eaubonne's writing circulated in a different and particular context." The context she had in mind was that of French politics, especially French feminism in the early 1970s, which is evident in the title of the collection in which the first English translation of d'Eaubonne's work appeared – *New French Feminisms* (Marks and De Courtivron 1980). Nonetheless, Sandilands's comment limits the possible influence of d'Eaubonne and risks suggesting that her work had little diffusion or meaning outside of France.

Others have produced alternative origin stories for eco/feminism, countering claims regarding the role of d'Eaubonne and pointing to the work of Ynestra King. According to Chaia Heller (1999, 53), "In 1978, the Institute for Social Ecology (ISE), which Bookchin co-founded in 1974, invited Ynestra King to develop what would become the first curriculum in a feminist approach to ecology, thus coining the term ecofeminism." Sturgeon and Sandilands also referred to the work of Ynestra King. Sturgeon (1997, 32) stated that "the first classes in ecofeminism in the United States were taught by King at ISE in 1978," and Sandilands (1999, 213) wrote that King "disputes this 'first' [d'Eaubonne's naming of ecofeminism], and was certainly active in the mid-1970s developing ecofeminism at the Institute for Social Ecology." Heller (1999, 65n20) explained her attention to narratives of eco/feminism, her decentring of d'Eaubonne, and her stress on the role of King by asserting that examining the lineage of the term is a way to explore the specific historical context in which ecofeminist theory and action emerged. Attempts to trace the ecofeminist movement itself back to d'Eaubonne obfuscate the historical continuity between ecofeminist curriculum and writing that emerged at the ISE (Institute for Social Ecology, Vermont) by King, and the wider context of the US New Left made up of activists involved in the radical feminist movement, the feminist peace movement, the anti-war movement, and the anti-nuclear movement.

Although some have reflected on the significance of d'Eaubonne's work and on the trail of footnotes and citations to which she has given rise, little evidence of critical reflexivity appears in academic eco/feminists' fascination with d'Eaubonne – a fascination that continued with her invitation to a 1998 eco/feminist conference held in Montana.[3] Perhaps inadvertently, a comment by Greta Gaard (1998, 13-14) on claims of d'Eaubonne's strong influence on eco/feminism in the United States is revealing:

> What none of these defenses of d'Eaubonne addresses is the different political implications of attributing the origin of the term (and, by implication, the movement and ideas behind it) to a lone, white, first world scholar – or to the "spontaneous combustion" of many women around the globe ... It would seem that the dispute over the lineage of "eco/feminism" has been to some extent a class war over whether the idea was born from a single woman labouring alone in the library or from many women labouring in the forests, the military bases, and the nuclear power plants. For ecofeminists who want to open up the movement in such a way that activists themselves will claim the term, a more populist origin for the word would seem to have more strategic appeal.

Disarmingly, Gaard (ibid., 14) concluded "fortunately it's also true" – that the term "eco/feminism" emerged from grassroots activism. Despite her intimation regarding the vested interests of academics, Gaard retreated from a critical examination of the stakes that white Western eco/feminists in the academy might have had in representing eco/feminism as an international grassroots movement. In addition, though Gaard may have been wary of crediting an individual such as d'Eaubonne with the emergence of eco/feminism, there is a hint that her stance was not only because of a desire for a collective and activist narrative of its origins. Commenting on the repeated references to d'Eaubonne, she noted that "the cumulative force of so many citations angered some and puzzled others, who felt certain that they had learned the term elsewhere." Writing of her first meeting with other eco/feminists at the National Women's Studies Association (US) convention in 1989, Gaard (ibid., 13, 3) recounted that "it was both humorous and frustrating to discover that so many of us, working alone in our separate parts of the country, thought we had invented these ideas ourselves."

Thus, though d'Eaubonne is as ubiquitous as Shiva and the Chipko Movement in examinations of eco/feminism, Shiva and Chipko dominated the limelight, whereas d'Eaubonne was often relegated to a brief footnote. The point of this discussion is not to provide a corrective, to defend or reassert the importance of d'Eaubonne's work, or to suggest the greater significance of another author. Instead, I pose the question of what is at issue in the citation of a French feminist in the naming of eco/feminism. There is no reason why d'Eaubonne's use of the term "ecofeminism" could not be seen as just another example of eco/feminism's internationalism were it not for the emphasis on the written, and on the ownership of ideas, in academia, compounded by the Westerncentrism of much academia. Clearly, *Le féminisme ou la mort* alone does not account for women's environmental activism of recent decades or even for the consolidation of a specifically eco/feminist theory. At the same time, as a French academic and hence a westerner, d'Eaubonne is not quite international enough with respect to Chipko and other movements of the global South. Thus, the academic convention of citing sources and origins acknowledges her as author of the term, but confining this information to a footnote or a passing mention means that she cannot be understood to author(ize) the movement. At the same time, the more overt references to Chipko provide proof of eco/feminist activism, but equally Chipko itself does not offer an intellectual foundation. Furthermore, as controversies over difference, race, and global sisterhood raged in feminism, eco/feminism faced growing criticisms over the citation, or appropriation, of Chipko, especially from feminists in the field of gender and development (see Moore 2008).

Eco/Feminist Responses to Critiques of Chipko

Celebratory accounts of the global, and of Chipko in particular, rapidly gave way to more cautious reflections. When discussing Chipko, these often focused on the anthologies mentioned earlier – both Plant (1989a) and Diamond and Orenstein (1990b) contain chapters by Shiva on Chipko as well as other chapters that refer to it. Sandilands (1999, 54) described the perceived limitations of these anthologies: "The inclusion of race was not especially analytical; it did not in most cases, suggest ways in which women may have different relations to particular ecological issues or problems, and it did not look deeply at the ways in which these traditions have themselves been lost or reconstructed in particular social contexts." She elaborated on the characterization of indigenous cultures as "somehow pure, somehow dissociable from what colonization has done to those different cultures and social practices. Also problematic was their general assumption that all 'women's' practices in nature are (at their core at least) benign, caring and respectful." Sandilands (ibid., 55, 56) concluded her discussion on a relieved note: "But (I'm very happy to say) this mode of discussion is no longer predominant in ecofeminist literatures that question racism and colonialism."

Others offered different resolutions to the apparent problems posed by these texts. One route was turning to research on women in the so-called developed world. Chris Cuomo (1998, 8-9) wrote that "while there is a tendency in Western ecofeminist theory to describe the work of rural Third World women as paradigmatic ecofeminist activism, one sees little effort (in the literature) to develop specific models that examine the politics of 'first world' megaconsumption on ecofeminist grounds." Cuomo's suggestion was taken up by Sherilyn MacGregor (2006, 128) in accounting for her own research with women in Ontario: "I am making the point that the experiences and ideas of urban-dwelling women in the overdeveloped world are as interesting and informative to ecofeminist thought as those of 'peasant' women in developing countries." Examining the activism of women in the West is understandable – after all, my own research is based in Clayoquot Sound – but this strategy brings problems of its own, not least erasing existing references to Western movements already commonly cited, including Greenham Common, the Seneca Women's Encampment, the Women's Pentagon Actions, and the strong anti-toxics movements in the United States. Many organizations and actions in the developed world have focused intensely on (over)consumption and its impact elsewhere. Such a resolution may also imply that (Western) eco/feminists can have nothing to say regarding women and the environment in the Third World. It also ignores ways in which Third World women can speak and be seen and heard by Western eco/feminists. Most significantly, it belies the myriad ways in which developed and underdeveloped, North and South, are intimately interconnected through the histories of colonialism and slavery, through the transference of goods and the effects of environmental degradation, and through the movements, both real and virtual, of activists. All of these connections are manifest in the work of many eco/feminist organizations. Some, such as Women Working for a Nuclear-Free and Independent Pacific, are alliances between women in the North and South. Thus, shifting one's focus from the South to the North requires careful attention to the specificities and contexts of histories and transnational connectivities.

Questions of race and indigeneity were central in Noël Sturgeon's *Ecofeminist Natures* (1997, 116), which identified Plant (1989b) and Diamond and Orenstein (1990b) as "the most prominent representatives of the diversity within ecofeminism." In focusing on discourses that "center on the idealization of 'indigenous' women as symbolic representatives of ecofeminism," Sturgeon (ibid., 113, 114) summed up some of the assumptions embedded in such uses of indigenous women: that non-industrialized cultures are seen as more ecological, as not the Western separation of nature and culture, and possibly as embodying more egalitarian gender relations. Nonetheless, one can revisit some of the criticisms of the anthologies. For instance, Plant's apparent reliance on Native American religious rituals in her own contribution to the collection, "The Circle Is Gathering," was disparaged by Sturgeon (ibid., 121). Yet this becomes more complicated when one attends to Plant's location in British Columbia – her discussion drew on specific First Nations cultures and her embeddedness in particular communities, rather than on generalized "Native Americans" as such. I do not wish to fuel Canadian exceptionalism or to ignore the fact that the indigenous people of North America have their own approach to national boundaries – my point is that Sturgeon's argument may rely on reading a specific First Nations iconography as a generalized indigeneity. Plant's ongoing commitment to bioregionalism may also contribute to the intentionality and specificity of her account and the imagery in her chapter and in her approach to the anthology. When *Healing the Wounds* was published in the United Kingdom, its cover was changed, possibly in recognition that it invoked a specific indigeneity that would not necessarily be understood in the UK context; the new cover featured a crystal that diffracts light in a rainbow of colours. Furthermore, critiques of the anthologies have tended to attribute agency to their editors but have paid much less attention to the intentionality of the contributors. One is left with the prospect that the women of colour whose essays appeared in these supposedly essentializing, universalist collections were naive romanticizers of their own cultures. How else could we make sense of the inclusion of black, Third World, and indigenous women who explicitly critique these kinds of discourses? It is not only a case of editors wanting to be inclusive, but also that authors may wish to be included in these collections, no matter how problematic they might seem. Perhaps this desire was in part due to the persistence of eco/feminism's orientation to global and transnational interconnections, as well as to environmentalism and spirituality. Thus, Carlassare's (1994) overview of eco/feminism was arguably the last moment that claims about diversity could easily be made. By the mid- to late 1990s, collective polyvocal texts such as anthologies and special issues of journals had been largely superseded by the monographs of white authors, most of whom were North Americans. As Greta Gaard (1998, 3) explained, she declined an invitation to write a monograph on eco/feminism, believing that it "would not do justice to such a multivocal grassroots movement." Gaard's first book on the subject was the anthology *Ecofeminism: Women, Animals, Nature* (1993b).[4] At the same time, Chipko and other movements slowly disappeared from the literature or became the focus of criticism, allowing authors to display sophistication by distancing themselves from them – this leads me to question whether responses to the assumptions of essentialism and universalism have been carefully and empirically worked through.

From Diversity to Typologies

Extending Sturgeon's genealogy and attending to other (apparently incommensurate) ways of accounting for eco/feminism is revealing. Descriptions that emphasize a diverse movement and certain iconic figures suggest an ongoing tension about the relationship between academia and activism, as embodied in the concerns about naming, authorship, knowledge production, and political action. This theme runs through mainstream feminism as well, where a tension between theory and activism has marked much recent history. What Robyn Wiegman (2008) termed the "theory-activism conundrum" might be understood as a defining feature of feminism, at least since the rise of women's studies and feminist theory in the academy; a commitment to this tricky nexus is part of what feminism is about. Eco/feminism is not unique in its struggle with this issue. In Sturgeon's (1997, 169, emphasis added) analysis of typologies, the stated aim is "to interrogate the anti-essentialist tangent within feminist theory (and, by extension, ecofeminist theory) that *created* a division between feminist theory and feminist activist practice." By extending Sturgeon's genealogical approach and exploring other sites of the naming and defining of eco/feminism, I suggest that the tension between theory and activism predated the turn to anti-essentialism, which simply exacerbated it. Even so, a key point to remember is that before typologizing came to dominate, theory, activism, spirituality, and much more co-existed, jostling alongside each other in accounts of eco/feminism. There may have been disagreement over who coined its name, but few viewed the name as unimportant or as anything other than a prize worth claiming and owning. At the same time, activism was widely perceived as a defining feature of eco/feminism. So typologizing might be better understood as a response to the increasing difficulty of managing the tensions between theory and activism, and as an attempt to avoid accusations of essentialism. Typologizing signals the moment when the insistence on the entanglement of theory, ritual, poetry, literature, food, activism, spirituality, and philosophy was abandoned, and when eco/feminism became defined through, and in relation to, the dualism of essentialism/anti-essentialism – rather than as an anti-dualistic philosophy and practice. Thus, typologizing marks a significant change in *how* we might know and understand eco/feminism.

Against diversity, typologizing is a process of categorization, an attempt to manage the range of meanings of eco/feminism. Yet, as Sturgeon very thoroughly illustrates, typologies of eco/feminism have proliferated. Their sheer variety might be perversely understood as a return of eco/feminism's diversity. It is possible to understand this as the failure of the binary of essentialism and anti-essentialism to hold eco/feminism in place, and the failure of theory and activism to neatly divide up, acting as a reminder that, though Sturgeon turns to the capacity of radical democratic movements to destabilize essentialisms (as I explore below), theory also has this ability to undo itself.

From Anti-Essentialism to Strategic Essentialism

In keeping with her critique of the idea that anti-essentialism defines the best eco/feminism, Sturgeon (1997, 9, 11) was keen to resist both essentialism and anti-essentialism, and she provocatively asked whether essentialism is always negative. Moving away from reductive notions of essentialism as itself an essence, she took up suggestions "to de-essentialize essentialism," "to differentiate what kinds of essentialism we are objecting to, and pay attention to the consequences." Indeed, her approach was to consider how essentialist rhetorics could, in certain situations, productively contribute to the creation of political communities. Specifically, she suggested that the essentialisms of eco/feminist movements could be better understood as "part of creating a shifting and strategic identification of the relation between 'women' and 'nature' that has political purposes." She (ibid., 169, 11) also argued that strategic essentialisms would continually be destabilized in the practise of radical democracy, citing "alliances, coalitions, networks, affinity groups, and consensus process decision making – that bring together different kinds of women for the purposes of political action" as the radically democratic structures that destabilize essentialist eco/feminist formulations.

In recognizing that activists were being villainized as essentialists, Sturgeon took a significant risk in nonetheless looking to activism as an important site of theory. She (ibid., 182-83, emphasis in original) deployed the concept of "direct theory" to counter reductive readings of essentialism in activism, rather seeing essentialisms in activist practice and rhetoric as "indicators of the *theorizing* going on *within* and *through* activist practice." Thus, a key intervention of her approach was to take both activism and the theorizing of activists seriously. Sturgeon (ibid., 11) suggested that "one way to understand solutions to what I see as a political stalemate between tropes of essentialism and anti-essentialism within feminism is to carefully theorize feminist activist practice and to see the theory in that practice." Sturgeon asserted that maintaining a connection with direct theory would offset the tendency of academics to separate themselves from the activist component of eco/feminism (and feminism). Sturgeon's work is distinctive because it tries to address the problematic effects of essentialism by challenging the categories of essentialist and anti-essentialist, and because of its attention to the value of empirical research on eco/feminist activism.

Nonetheless, it is debatable whether strategic essentialism can bear the weight of Sturgeon's expectations. While some feminists did initially suggest the potential of strategic essentialism, most have now revisited any enthusiasms they once held for it. Like essentialism, it has a fraught and contested history. Stephen Heath's (1978, 99) suggestion that "the risk of essence may have to be taken" was also echoed by many others (Jardine 1987; Spivak 1987; de Lauretis 1989; Fuss 1989; Schor 1994). Gayatri Spivak (1987, 205) is particularly associated with strategic essentialism, having once called for the "strategic use of essentialism in a scrupulously visible political interest." Yet Spivak (1994, 154) subsequently expressed her surprise and irritation about the frequency with which her suggestion was taken up, and she later reconsidered it. Andermahr, Lovell, and Wolkowitz (2000, 82) mused that strategic essentialism may succeed only in displacing the confrontation between itself and essentialism per se: "My essentialism is strategic and therefore virtuous; yours however is fundamentalist and vicious." Fuss (1989, 19) reflected that strategies are always unpredictable and undecidable: "Depending on the historical moment and the cultural context, a strategy can be 'radically revolutionary or deconstructive' or it can be 'dangerously reactive.'" The implication of calling into question whether essentialisms are always reactive, and constructionist positions always radical, is that "the radicality or conservatism of essentialism depends to a significant degree, on who is utilizing it, how it is deployed, and where its effects are concentrated" (Fuss 1989, 20). Though Fuss was speaking of essentialism here, her point also applies to strategic essentialism and anti-essentialism. To a certain extent, Sturgeon's approach – to examine activism and not to assume essentialism, or at least not a regressive essentialism – mirrors that of Fuss, but routinely applying strategic essentialism to activisms in all their diversity undermines her method. Nor is it clear how direct theory is being understood, whether it is mobilized by Sturgeon or by activists. Thus, though Sturgeon assumes that strategic essentialisms are "almost always" destabilized by radical democratic processes, the grounds for her optimism are not entirely clear, and her "almost always" implies that some strategic essentialisms will remain untouched. What if, in fact, they are mobilized to react to and shore up the challenges of anti-essentialism in practice? (An instance of this, which occurred at the Clayoquot Peace Camp, will be discussed in Chapter 5.) Sturgeon's recuperation of eco/feminist activism relies on the reduction of any number of diverse instances of activism to a single moniker, strategic essentialism. Despite her careful attention to activism, her approach risks a rescue logic by theoretical pronouncement of strategic essentialism rather than through empirical examination. How such diverse practices can be identified as strategic essentialism is not clear. After all, the terms "essentialism" and "strategic essentialism" are not much used by activists, or by many people, in their daily lives. As Jana Sawicki (1991, 26) pointed out, "Strategic essentialism is a strategy – not a theory." She wrote that "the question whether some forms of resistance are more effective than others is a matter of social and historical investigation and not a matter of prior theoretical pronouncement." Sawicki's reflections point toward the *methodological* implications of a reliance on any essentialism, strategic or otherwise. Therefore, though Sturgeon's work offers one of the best models for eco/feminism, her arguments rely heavily on the power of strategic essentialism and on radical democracy to undo essentialist knots of all kinds. Indeed, I would argue that strategic essentialism offers no resolution at all. As much as essentialism, it may foreclose the possibilities of theorizing. Ultimately, the supposed problem of essentialism cannot easily be resolved by anti-essentialism or strategic essentialism; it demands attention to the empirical and to methodology. As Spivak (1994, 160) suggested, "Anti-essentialism is a way of really not doing one's homework."

Yet, in the desire to prove that eco/feminism is not essentialist, that it is resolutely anti-essentialist or can be recuperated as strategically essentialist, eco/feminists have also implicitly conceded that essentialism has indeed constituted an appropriate way to theorize eco/feminism and women and nature. I argue that the shift to typologizing – by eco/feminists themselves – marked the moment when eco/feminist academics succumbed to feminist narratives of essentialism and gave up their earlier optimism regarding eco/feminism and its radical diversity. This is suggested in the introduction to the anthology *Ecofeminism and Globalization* (Eaton and Lorentzen 2003, 5): "Although it is considerably developed in both popular movements and academic discourse, ecofeminism remains largely a theoretical conversation ... While there are many grassroots activist women's organisations resisting the negative effects of globalisation, these activities do not provide the primary data for ecofeminist discourse." Eco/feminist academics began to cede much of eco/feminism's radical potential to essentialism – as can be seen in the quotes that open this chapter. After all, Sturgeon began her chapter on typologies with the puzzle of feminists' apparent rejection of eco/feminism and a sense of their failure to recognize eco/feminists as feminists, all achieved through their accusation of essentialism. Sturgeon's (1997, 168) plaintive question – "What has the ecofeminist movement done or not done to deserve its exclusion from certain feminist circles?" – is indicative of some eco/feminists' bewilderment over the failure of feminists to recognize their feminism.

In "What's in a Name?" the concluding chapter of *Ecofeminist Natures,* Sturgeon was more explicit about this. She examined the related practices of creating new terms for eco/feminism (such as ecological feminism) and of generating typologies of various kinds of eco/feminism (radical ecofeminism, socialist ecofeminism). This resulted in differentiating "anti-essentialist" and "essentialist" eco/feminisms, which were mapped onto post-structuralist/academic feminisms and activist and spiritual feminisms – that is, "good" and "bad" eco/feminisms. Importantly for my account, Sturgeon (ibid., 173) identified "two common results of the practice of typologising: the invisibility of women of colour and the creation of a divide between feminist theory and feminist activism," thereby demonstrating how Chipko is doubly disadvantaged in such accounts of eco/feminism.

From Chipko to Clayoquot and Back Again

With this context in mind, I return to Clayoquot to explore the campaigning of the FOCS. I take as a departure point a moment in my interview with Fireweed, one of the women from the camp, when she described a conference that she had recently attended in Victoria.[5] Following a talk from Vandana Shiva, FOCS director Valerie Langer stood up and told Shiva that the work of the Chipko Movement was being continued in Clayoquot. In doing this, Langer was not alone in publicly connecting Chipko with Clayoquot. The documentary *Fury for the Sound: The Women at Clayoquot,* which was widely shown on Canadian television, opens with a clip of Shiva (Wine 1997). Chipko was no doubt invoked in the many conversations and writings about the campaign. Yet, in the context of eco/feminist controversies over difference, race, and the international, it has become difficult to read Langer's claim and its approving repetition by Fireweed as anything other than yet more damning evidence of eco/feminism's ongoing essentialism, universalism, racism, neo-colonialism, imperialism, romanticization of subsistence practices, ignorance, naïveté, and so on; there are a range of pejoratives to choose from. However, this is precisely what I want to attempt, through posing a number of questions: how has it become so easy to assume that we understand the meaning of Langer's statement? How has it become so easy to produce and proliferate such definitive truths about eco/feminism, Shiva, Chipko, and the international more generally, to the exclusion of other stories? What work have such characterizations done? What purposes have they served? What are their, possibly unintended, consequences? How might Langer's statement be read differently and what would the implications of this be?

What if, rather than assuming essentialism on the basis of a mention of Chipko, we admitted the possibility that Langer (and the FOCS) possessed extensive *knowledges,* particularly of the multiple entanglements of the local and the international (see also Magnusson and Shaw 2003b)? In too many ways to enumerate here, some of this knowledge was visible in the campaign; for instance, it was no accident that I first came across Clayoquot in the London office of the Women's Environmental Network (WEN). At that time, WEN was the only European environmental organization that had a temperate rainforest campaign, and it was an initial contact point for the FOCS when it wanted to bring its campaign to Europe as it followed the path of the trees, exported from British Columbia in the form of timber and wood pulp. In concentrating on how women might bear the brunt of some environmental problems, as well as how women in the North were implicated in the overconsumption of disposable paper products that originated elsewhere, as in British Columbia, WEN's activism was crucial to the FOCS. Its focus on the connections between breast cancer, dioxins, land fill, bleached toilet rolls and sanitary towels, and on the clear-cutting of old-growth forest in British Columbia might meet the concerns voiced by Cuomo (1998, 8-9) regarding the lack of attention to First World megaconsumption.

At the same time, as I argue in Chapter 4, the FOCS (and other BC environmental groups) suggested that the emphasis on the destruction of the Amazon rainforest allowed countries such as Canada to blame the developing world for overpopulation and environmental devastation, even as they themselves were responsible for incredible environmental damage. Thus, the Brazil of the North campaign formed part of efforts to hold Canada to account for the global implications of its activities: "Canada is the Brazil of the North. Brazil is losing one acre of forest every nine seconds. We're losing one acre every twelve seconds."[6] Internationally, other environmental organizations responded to this approach, and protests were held outside Canadian embassies in New York, London, Germany, and Japan, countries where BC wood pulp was used to make toilet paper, newsprint, and disposable chopsticks. This campaign to hold the Canadian government to account by shaming it internationally did not go unheeded. When the FOCS took its campaign to Europe, and a European boycott of BC forest products was threatened, the pro-logging lobby asserted that FOCS members were traitors who were guilty of treason (Rowell 1996; see also Moore 2003).

Global connections also have histories. In 1914, H.R. MacMillan, who would eventually establish MacMillan Bloedel, the company responsible for logging in Clayoquot Sound, travelled to India to develop a significant export market for BC timber. As a result of his efforts, trees from British Columbia eventually made their way to India, where the lumber industry was struggling to meet the demands of British colonists to extend railways, requiring a seemingly endless supply of wooden sleepers. But more than timber crossed international borders; the nascent field of forestry science criss-crossed the world to be tested in British colonies such as India before being applied in other places, such as Canada (Drushka 1995; Tsing 1997). Much later, Vandana Shiva would come from India to study in Canada and would keep returning, including for conferences on feminist theory and praxis such as the one where Fireweed heard her speak.

I could go on, weaving the webs of connections that are really what is essential to eco/feminism. This process calls for an openness to alternative stories and knowledges, and a willingness to risk the accusation of universalism. My point here is that Langer's "We are carrying on the work of the Chipko Movement" cannot even straightforwardly be assimilated into an account of white women's benevolence, philanthropy, or virtue ethics. The FOCS had no intention of raising funds for the Chipko women or of visiting India to help them hug trees; rather, it focused on the recognition that women in Canada, in Clayoquot, were implicated in the global trade in trees and timber (and forest science), that the most appropriate locus of action might be its own backyard (to reverse the accusation of NIMBYism, which is often applied to environmental activists), and that appropriate action might entail challenging the Canadian government's complicity in such global trade and global environmental devastation.

We might understand the invocations of Chipko not as signalling a universalized essential femininity but as a genealogy of women "as a motivated, and motivating practice" (Haran and Moore 2008), as re-visioning a community of eco/feminists. To genealogy as re-visioning we could also add the work of "reclaiming." Here I echo Isabelle Stengers (2008, 41-42), who explores reclaiming Shiva's work specifically, against what she terms "essentialist hunting" (a reference to the witch hunts). She states that the relevant question is "Can we separate Vandana's force – which produces her ability to struggle – from those seemingly 'essentialist' grounds? And the challenge would be learning to ... hesitate about our own conditions of thought." Drawing on the work of US Wiccan eco/feminist Starhawk, who came to the camp and was arrested there, Stengers (ibid., 58) expands on her practice of reclaiming and its capacity to make us hesitate:

> Reclaiming is an adventure, both empirical and pragmatic, because it does not primarily mean taking back what was confiscated, but rather learning what it takes to inhabit again what was devastated. Reclaiming indeed associates irreducibly "to heal," "to reappropriate," "to learn/teach again," "to struggle," to "become able to restore life where it was poisoned," and it demands that we learn how to do it for each zone of devastation, each zone of the earth, of our collective practices and of our experience.

These kinds of understanding of genealogical practices are also suggested by Anna Lowenhaupt Tsing's work, which is useful not least because of her focus on the destruction of forests, in her case in Indonesia. In *Friction: An Ethnography of Global Connection,* Tsing (2005) turned to list making in her discussion of collaborations and specifically of biodiversity assessment as a multicultural exercise. At the insistence of a friend and mentor in the Meratus Mountains of Indonesia, she helped create a list of the plants and animals of Borneo. Noting the importance of species lists in making conservation claims and recognizing that these lists enable "us to discover variety and to appreciate dynamics," Tsing (2005, 162) offered her list as "a motivated set of translations and not a simple addition to either universal or local knowledge culture." She reflected that "list-making is eclectic to the extent that it draws on multiple, fragmentary sources. To acknowledge this eclecticism allows us to admire its creative use of limited materials, rather than to grasp only for scope. It allows us to imagine the list within historically changing conversations, rather than as transcendental knowledge." Thus, Tsing's account of list making is profoundly genealogical.

Understanding list making as a genealogical knowledge practice is instructive in re-visioning and reclaiming eco/feminist activisms, offering ways of reading these not as essentialist or universalist, but rather as consisting of linked specificities. The list includes Chipko, the Kenyan Greenbelt Movement, the Women's Pentagon Actions, the Narmada Valley dams campaign, the Love Canal Homeowners campaign against toxic waste dumping, the women who organized around Chernobyl, Bhopal, nuclear testing in the Pacific, Japanese women's consumer movements, Greenham Common, the Women's Environmental Network – and now Clayoquot Sound. In the absence of any agreed definition, the list provides an introduction to what eco/feminism is about – connections between the oppression of women and the domination of nature. It describes eco/feminism without employing generalizations that would have far too many exceptions. It hints at the complexity of women's relationships with nature and their environments. It demonstrates that there is no single eco/feminism and that eco/feminism is not merely a fantasy of feminist academics. In the changing context of anxiety about the essentialisms of activism and the universalisms of imagined global sisterhood, the diverse locations of these activisms are as significant as the actions themselves, offering a counternarrative to the end of global sisterhood and of feminism to anyone who wishes to listen. The diversity initially valorized by eco/feminists was not only activism or international activism, but also a diversity of forms, as is clear in the remarks of Carlassare (1994, 220-21) earlier and in the introductions to the eco/feminist anthologies. Diverse genres, forms of knowledge, and knowledge practices were included. Yet, in the understandable concern regarding the predominance of Third World activism and Native American spiritualities, this attention has been lost.

Against the deadening repetition of the end of feminism, against the typologies of eco/feminism that sought to purify theories of essentialism and activism, and against the progress narratives of certain feminist histories, some women were briefly, excitedly, repeating the names of places around the world and passing them on to each other, not necessarily as a manifestation of a universal essentialized womanhood, but as a reminder of the work that still needed to be done, an invocation to action. I understand the practice of listing eco/feminist activisms as a kind of ritual, a performative recitation in Butler's sense, which insisted on the persistence of feminist activism and hoped to bring eco/feminism into being, while simultaneously warding off the challenges of those who denied the possibilities of connection. This list has many beginnings but no single ending. The challenge for eco/feminist academics was to determine how to articulate ourselves into this community, genealogy, of eco/feminists and to decide what we must do (to) ourselves. At the very least, the work for eco/feminist academics might have been – and might yet still be – to articulate the complicated, messy tensions and frictions in the unfinished, open-ended listing, genealogy of eco/feminism, which theorizes and enacts, which traces histories, records the present, and conjures imagined communities of eco/feminists, a genealogy that I continue to recite, reclaim, and re-vision: ... the Chipko Movement, the Kenyan Greenbelt Movement, the Women's Pentagon Actions, the campaign against the Narmada Valley Dams, the Love Canal Homeowners campaign against toxic waste dumping, the women who organized around Chernobyl, Bhopal, nuclear testing in the Pacific, Japanese women's consumer movements, Greenham Common, the Women's Environmental Network, Clayoquot Sound ...

3

Eco/Feminism and the Question of Nature

> But I often return to that moment, lying on the ground, surrounded by all these wonderful women – a chance grouping that in hindsight seems somewhat fantastic – and thinking: What is this really about? Why did these amazing women get together to do this rather bizarre action? What kind of movement is this?
>
> In some ways, the Surrogate Others [affinity group] were the beginnings of this particular book, this particular effort to understand the movement, that in my personal history, was tied historically to a movement now often called U.S. ecofeminism. Were the Surrogate Others "Cyborgs for Earthly Survival," as the political button invented by Elizabeth Bird proclaimed? Or were we part of an ecofeminist movement, that as Cecile Jackson has put it, is "ethnocentric, essentialist, blind to class, ethnicity and other differentiating cleavages, ahistorical and neglects the material sphere?"
>
> – NOËL STURGEON, *ECOFEMINIST NATURES* (1997)

> Feminist environmentalism has become bogged down in tiresome "pro-/anti-ecofeminist" reprises: Is the earth our mother? Are women closer to nature than men? Should they be? Should we decry the cultural association of women with earthy nature as a patriarchal contrivance or celebrate it as a privileged positionality? Is ecofeminism even feminist (or is it a complicity with the patriarchy that would inexorably bind women to nature, to the disadvantage of both)? Is the disparagement of ecofeminism by (many) academics reductive elitism (and a complicity with the patriarchal appeal to masculinist reason that would separate women from a meaningful relationship with their sacred earth origins)? We have interrogated these questions to the point of exhaustion, and well past the point of diminishing intellectual and political returns.
>
> – JONI SEAGER, "RACHEL CARSON DIED OF BREAST CANCER" (2003)

> Ecological feminism has a history of deliberating such difficult questions as: Are women more "naturally" connected to nature than men? Do women's gendered roles and experiences give them unique insight into human-nature relationships? Why is it that women around the world seem to demonstrate relatively more concern for the quality of their environments than men?
>
> – SHERILYN MacGREGOR, *BEYOND MOTHERING EARTH* (2006)

Essential(ist) Entanglements

As these epigraphs suggest, eco/feminism has often been understood as addressing such questions as whether women are closer than men to nature, and relatedly, whether women are more concerned about the environment than men. These questions have been deeply entangled with the seemingly unavoidable problematic of whether affirming a relationship between women and nature indelibly marks one as an avowed essentialist or whether claiming an anti-essentialist position, to be a "cyborg for earthly survival," is possible. Like Seager, I am not convinced that pursuing these questions has been fruitful. However, unlike Seager, I am not sure that eco/feminists can easily proceed without engaging with the subject of essentialism.

Otherwise, it seems that the question – or the assumption and accusation – of essentialism remains to haunt eco/feminism. Over a relatively short time, open-ended questions of women and nature, feminism and environmentalism, collapsed into a matter of attributing essentialism or anti-essentialism, leaving a cramped space for eco/feminism. For instance, in discussing the campaign against nuclear testing at the Nevada Test Site, the subject of the first epigraph above, Sturgeon (1997, 3) initially posed the open and generative questions "What is this really about? Why did these amazing women get together to do this rather bizarre action? What kind of movement is this?" Yet these empirical questions quickly reduced to the binary options of either "cyborgs for earthly survival" or "ethnocentric essentialists." Yet it seems to me that, like feminist and eco/feminist theorists and academics, activists rarely identify themselves as essentialist; nor do they tend to use the language of anti-essentialism or, indeed, to see themselves as cyborgs for earthly survival. The affinity group at the Nevada Test Site was rather unusual, including Donna Haraway, for instance (Sturgeon 1997, 2; see also Haraway 1992). A *question* of nature, of women's nature specifically, became an *accusation* of essentialism. It sometimes seems that eco/feminists, particularly from the 1990s onward, have (had to) become so enmeshed in defending against the charge of essentialism that little energy remains for articulating and working through – or even remembering – alternative accounts of eco/feminism, or of women and nature. Yet I doubted that asking the women at Clayoquot whether they felt closer than men to nature would help explain why they ended up there or what being there meant to them. Neither would it tell us much about what happened at Clayoquot or about feminist and eco/feminist activism of any kind. Rather, this chapter seeks to find ways to open up the apparently closed question of women and nature.

Yet eco/feminist academics were not alone in confronting the dilemma of the empirical, of women's experience, and of essentialism – this is very well-trodden ground for feminist academics. In 1997, Beverley Skeggs (1997, 2) noted that "there has been a marked tendency in recent years to move away from talking and listening to those outside of academia," and in her own work she sought to demonstrate "how theory can be radically transformed if others are let in on the conversations." Essentialism has not only been a challenge for eco/feminists, but has been bound up with broader dilemmas in feminism, about experience and foundationalist politics. Throughout this book, I suggest that eco/feminist dilemmas are often feminist dilemmas writ large. My research is, therefore, in part an exploration of how eco/feminist and feminist theory can be transformed if others – women environmental activists and nature – are let in on the conversations. Given the persistent notion that eco/feminism is about women as close(r) to nature, and that this is irreducibly essentialist, in this chapter I outline a number of shifts required in order to foreground some important distinctions in how and why I went about my research in the way that I did, in trying to find ways of asking the question of women and nature afresh.

From Essentialism and Anti-Essentialism to "Many, Many Words"

In fact many feminists have been rethinking essentialism and its critique, so I begin by taking up their work. A number of crucial shifts have occurred in this work: a questioning of whether essentialism is always bad; an acceptance of the necessity and inevitability of essentialisms; and a rejection of anti- or strategic essentialism as offering any substantial way forward. These reflections have led towards a careful examination of how essentialism has been invoked and to what ends. Arguably, eco/feminists have not always drawn on this useful rethinking of essentialism *within* feminism, neither have they sought to explicitly intervene in this debate – this is what I wish to pursue.

Feminists once closely associated with a critical interrogation of essentialism have long since distanced themselves from their initial unfavourable assessments of essentialism (Spivak 1989; Butler 1990; de Lauretis 1989, 1990, 1994; Butler and Scott 1992; Phelan 1992; Butler 1993; V. Bell 1999). Teresa de Lauretis (1989, 3) reflected that "many who, like myself, have been involved with feminist critical theory for some time, who did use the term, initially as a serious critical concept, have grown impatient with this word – essentialism – time and time again repeated with its reductive ring, its self-righteous tone of superiority; its contempt for 'them' – those guilty of it." Ultimately, the insistence on anti-essentialism was identified as being as insidious as the critique of essentialism, resulting in "the policing of feminism by the shock troops of anti-essentialism" (Schor 1994, vii). As Spivak (1994, 156) explained, "What I am very conscious of is how anti-essentialism, really more than essentialism, is allowing women to call names and congratulate themselves." The critique of essentialism is no longer, if indeed it ever was, a meaningful, engaged critique, but rather became an accusation of not being a proper feminist. Whereas anti-essentialism was initially posed as the definitive answer to the problem of essentialism, in due course it was demonstrated that anti-essentialism relied on essentialisms and exclusions of its own (Fuss 1989; Butler and Scott 1992; Ahmed 1998).

Spivak (1994, 157) suggested that "the critique of essentialism should not be seen as being critical in the colloquial, Anglo-American sense of being adversely inclined, but as a critique in the very strong European philosophical sense, that is to say, as an acknowledgement of its usefulness." Unfortunately, this sense of "critique" was not widely taken up, with profound implications for the development of eco/feminism. In seeking a more generative response to the challenges of essentialism, Diana Fuss (1989, xi, emphasis in original) argued that

> in and of itself, essentialism is neither good nor bad, progressive nor reactionary, beneficial nor dangerous. The question we should be asking is not "is this text essentialist (and therefore bad)?" but rather, "if this text is essentialist, *what motivates its deployment?*" How does the sign "essence" circulate in various contemporary critical debates? Where, how, and why is it invoked? What are its political and textual effects? These, to me, pose the more interesting and ultimately the more difficult questions.

Like Fuss, Spivak stressed the importance of tracing the movement of essentialism and the uses to which it has been put. As well as calling for critical theorists to take the risk of essentialism, she asked how we might rethink it. For Spivak (1994, 156), the proper goal of essentialist critique was not the exposure of error but the persistent examination of the production of truths: "If I understand deconstruction, deconstruction is not the exposure of error, certainly not other people's error. The critique in deconstruction, the most serious critique in deconstruction, is the critique of something that is extremely useful, something without which we cannot do anything. That should be the approach to how we are essentialists." This feminist rethinking of essentialism is particularly provocative for reimagining eco/feminism. Rather than providing a theoretical resolution to the supposed problem of essentialism, it allows us to examine how essentialism is invoked in the production of various truths, such as the "truth" that eco/feminism is essentialist.

The shift from taking essentialism as inherently bad to tracing how it is used and to what ends was productive in the previous chapter in following how (academic) eco/feminists have negotiated essentialism. At the same time, tracing the use of essentialism offers only a limited agenda for eco/

feminist research. Spivak (1994, 171) proposed a more expanded research agenda in her attention to the methodological implications of the critique of essentialism. In moving away from "theorizing essentialism," she pointed to the confusion of, and conflation of, being empirical and the pursuit of essences: "The general subject of essence is not a good basis for investigation. This is rather different from being empirical." Essentialism, then, is understood as a problem of reading: "It seems to me that to be empirical in this way would be a much greater challenge, require much harder work, would undermine ... would make people read different things." Hence, she suggested that "if one begins to see what words one could use in place of essentialism or essence many, many words would come in." Thus, Spivak's work offers a way of countering the critique of essentialism through the possibilities of an empirical agenda for eco/feminism.

My next turn is to examine some of these many, many words that offer alternatives to essentialism and open up imaginative possibilities for eco/feminists include Haraway's cyborgs and naturecultures, Butler's performativity and genealogy, and Starhawk's goddess and spiral dance.

Eco/Feminist Figurations: The Spiral Dance of Cyborgs and Goddesses

Donna Haraway, who has done much to challenge conventional ways of thinking about nature, has certainly "done her homework" and introduced "many, many words" in place of essentialism. In particular, she has employed figures, such as the cyborg, to perform this work. Key to Haraway's (1992, 298) project is her proposal that "nature is made, but not entirely by humans; it is a co-construction among humans and non-humans," and much of her writing works through the subsequent implications. As part of her commitment to resisting dualisms, essentialism, and foundationalism, she has introduced intense compound neologisms such as "naturecultures." Her work is compelling – especially for those who are interested in the question of women and nature. Haraway directly addressed the question of nature in "The Promises of Monsters" (ibid., 296):

> So, nature is not a physical place to which one can go, nor a treasure to fence in or bank, nor an essence to be saved or violated. Nature is not hidden and so does not need to be unveiled. Nature is not a text to be read in the codes of mathematics and biomedicine. It is not the "other" who offers origin, replenishment, and service. Neither mother, nurse, nor slave, nature is not matrix, resource, or tool for the reproduction of man.

> Nature is, however, a topos, a place, in the sense of a rhetorician's place or topic for consideration of common themes; nature is, strictly, a commonplace. We turn to this topic to order our discourse, to compose our memory. As a topic in this sense, nature also reminds us that in seventeenth-century English the "topick gods" were the local gods, the gods specific to places and peoples. We need these spirits, rhetorically if we can't have them any other way. We need them in order to reinhabit, precisely, common places – locations that are widely shared, inescapably local, worldly, enspirited; i.e., topical. In this sense, nature is the place to rebuild public culture.
>
> Nature is also a tropos, a trope. It is figure, construction, artifact, movement, displacement. Nature cannot pre-exist its construction ... Nature is a topic of public discourse on which much turns, even the earth.

In her "Manifesto for Cyborgs," Haraway (1985) famously turned to the cyborg as a generative figure, later a material-semiotic entity, for resisting dualisms and thinking through the breaching of boundaries – of human and machine, machine and organism, animal and human. For Haraway (ibid., 92), the cyborg offered a counter to organic holism, especially in radical feminism:

> American radical feminists like Susan Griffin, Audre Lorde, and Adrienne Rich have profoundly affected our political imagination – and perhaps restricted too much what we allow as a friendly body and political language. They insist on the organic as opposed to the technical. But their symbolic systems and the related positions of ecofeminism and feminist paganism, replete with organicisms, can only be understood in Sandoval's terms as oppositional ideologies fitting the late twentieth century. They would simply bewilder anyone not preoccupied with the machines and consciousness of late capitalism. In that sense they are part of the cyborg world.

Griffin, Lorde, and Rich might all be understood as significant influences in the emergence of eco/feminism, and given Haraway's remarks above, some might be surprised to learn that she too has influenced many eco/feminists. So I suggest it is worth remembering the spiral dance – it is as much the case that Haraway's socialist feminist cyborg makes little sense outside of certain aspects of late-twentieth-century (and early-twenty-first-century) feminism, as the goddess or the organic. While Haraway insists on the breaching of boundaries, it is notable that the cyborg and goddess are bound together in a spiral dance – but their boundaries, and the boundaries of the sacred and the technological or the secular, are not breached here. It is not clear to me that Griffin, Lorde, and Rich do insist on the organic as opposed to the technical, or Haraway's version of organic holism in any case. They could also be read as offering accounts of the kind of interconnection that Haraway is certainly interested in. Rather, they may have had a profound awareness of the challenges of the work that needed to be done. Lorde's (1984) famous dictum – "The master's tools will never dismantle the master's house" – may have some bearing on the cyborg. The unanticipated and unprecedented take-up of the figure of the cyborg appears, on the one hand, as testament to Haraway's skill and imagination. On the other hand, it can perhaps also be understood as an overestimation of how terrifying the breaching of human and technology would be and an underestimation of how open to professionalization the cyborg was. After a forceful career, the cyborg is now being retired, although the need for the organic and the goddess, and much more, appears to remain intense. Haraway's turn to companion species, specifically the dog, can be understood as part of her continuing sense of responsibility for the extraordinary adoption of the cyborg, and the cyborg's unintended consequences, in her recognition that "as an oppositional figure the cyborg has a rather short half-life" (Haraway 2003, 4; Haraway 2004, 326). Rather than going to the dogs here, I turn to the figure of the goddess, who appears in Haraway's work as well, if perhaps as a disavowed figure and certainly not one as elaborate(d) as the cyborg or the dog.

So (and for those for whom dogs are not enough), I turn to Haraway's now (in)famous closing provocation in "A Manifesto for Cyborgs" (1985, 101): "Though both are bound in the spiral dance, I would rather be a cyborg than a goddess." Given Haraway's sense of humour and her irony, it is difficult to take this remark too seriously, or alternatively, there is a risk of not taking it seriously enough. Haraway both recognizes and demonstrates the power of dualisms, echoing and evidencing Spivak's account of the necessity – and danger – of essentialism and the difficulty of writing without invoking an "other" for the cyborg, here, the goddess. Haraway's declaration might be understood as playing with a return to dualisms. However, the spiral dance is key here. Not so much opposite sides of a dualism, the cyborg and the goddess are more usefully understood as different naturecultures, different material-semiotic entities, different moments in the ritual dance of naturecultures. The operative part of Haraway's sentence might be "I'd rather," which gestures to her ambivalent relationship with religion, and spirituality, even as earlier in her essay, she (ibid., 71) left open the possibilities of the spiral dance: "Ironically, it might be the unnatural cyborg women making chips in Asia and spiral dancing in Santa Rita jail whose constructed unities will guide effective oppositional strategies." As a final sentence for a manifesto, though it appears to be proclamatory in tone, it might also (now) be read as confessional, not least given Haraway's stated Catholic imaginary.

Although Haraway does not (yet) pursue the figure of the goddess, others have found the figure generative.[1] Haraway herself appeared to allow the cyborg a latitude and capacity that she did not accord the goddess. We might understand Haraway's (2004, 328) predilection for the cyborg rather than the goddess as echoing her preference for coyote rather than Spiderwoman: "I chose coyote and not, for example, Spiderwoman, because of the already overdetermined feminist appropriations of the latter." Though Haraway's account of the cyborg was surely a challenge, both for her and for her readers, I also find it useful to pose questions regarding Spiderwoman and the goddess. What would it mean to elaborate Spiderwoman as a fully articulated figure, as a material-semiotic entity? And similarly, what would it mean to elaborate the goddess as a fully articulated figure, as a material-semiotic entity – who otherwise remains "reduced" to organicism and holism in Haraway's account? I suggest that neither Spiderwoman nor the goddess have been appropriated enough, precisely *because* of *some* feminist overdeterminations of them. Whereas Haraway offered a detailed and generous account of the cyborg, I cannot help but wonder how eco/feminist history would have been different had she written as carefully of the goddess.

In referring to the spiral dance, Haraway was surely obliquely alluding to Starhawk's *The Spiral Dance* (1999), though curiously, given her famed attention to citational practice, she did not mention the book by name.[2] An eco/feminist, Wiccan, and activist, Starhawk created the ritual of the spiral dance in 1979, which she and others performed in San Francisco during Samhain to launch the book; it has been performed annually in San Francisco ever since. The cyborg and the spiral dance were both unprecedented and unimagined successes, taking on lives and afterlives not at all anticipated by their progenitors. The account of the goddess offered by Starhawk in *The Spiral Dance* also suggests that even if we might all be cyborgs, we are not necessarily precluded from being goddesses too. Starhawk (ibid., 10) specified not that we *are* goddesses, but that the goddess is "embodied, that we are each a manifestation of the living being of earth, that nature, culture, and life in all their diversity are sacred." In her novel *The Fifth Sacred Thing,* which was published in 1993, the same year as the Clayoquot camp, Starhawk offered a further account of the goddess, perhaps in an oblique response to Haraway. Set in North America in 2048, after a holocaust, the novel focuses on a community trying to rebuild itself in the ruins of San Francisco while facing the threat of war from the Millennialists, fundamentalist Christians in the City of Angels, once Los Angeles. It opens as ninety-eight-year-old Maya Greenwood reflects on ritual celebrations during this year's Day of the Reaper, the twentieth anniversary of the Uprising against the Stewards, who cancelled elections and imposed martial law:

> Maya could feel the earth under her, alive like a beating heart. Or perhaps, she thought, I'm feeling my own throbbing feet? Still, it was good, at the place of the dead, to acknowledge that One to whom she had pledged herself long ago, the aliveness at the heart of things, the ever-turning wheel of birth, growth, death and regeneration. It had occurred to Maya lately that calling *that* the Goddess, even though she had fought for the term all her life, was – what? Not so much a metaphor, more in the nature of an inside joke. (ibid., 13-14)

Thus, the goddess is not a metaphor, but rather a joke, an inside joke, which might not be so obvious to those who do not dance the spiral dance and who might (mis)take her too seriously; like the cyborg, the goddess has a perverse sense of humour. She might best be understood as queer kin to the cyborg, another material-semiotic entity – or perhaps a material-semiotic-spiritual entity.

So, although it may seem surprising to some that eco/feminists draw on Haraway's work, such a response constitutes a reductive reading of both it and eco/feminism, not least because of the difficulty of separating eco/feminism and feminism. Here Haraway herself is a case in point, as a feminist who has sometimes explicitly taken up the work of eco/feminists and has rejected simplistic readings that reduce eco/feminism to essentialism. On the back cover of Noël Sturgeon's *Ecofeminist Natures* (1997), Haraway insisted that "(ecofeminism) must not be stereotyped as an essentialist dogma, frozen at one caricatured historical moment." Elsewhere, she (1992, 297) asked, "Haven't eco-feminists and other multicultural and intercultural radicals begun to convince us that nature is precisely not to be seen in the guise of the Eurocentric productionism and anthropocentrism that have threatened to reproduce, literally, all the world in the deadly image of the Same?"[3]

Much of Haraway's account of naturecultures could apply as easily to the goddess as to the cyborg. The following passages are exemplary:

> All actors become who they are in the dance of relating, not from scratch, not ex nihilo, but full of the patterns of their sometimes joined, sometimes separate heritages both before and lateral to this encounter. (Haraway 2008, 25)

> Nature for us is made, as both fiction and fact. If organisms are natural objects, it is crucial to remember that organisms are not born; they are made in world-changing technoscientific practices by particular collective actors in particular times and places. (Haraway 2004, 65)

> The actors are not all "us." If the world exists for us as "nature," this designates a kind of relationship, an achievement among many actors, not all of them human, not all of them organic, not all of them technological. In its scientific embodiments as well as in other forms nature is made, but not entirely by humans; it is a co-construction among humans and nonhumans. (ibid., 66)

Perhaps to naturecultures we might add cyborggoddesses, genealogically linked with Lati Mani's *SacredSecular* (Mani 2009). Arguably, decontextualized from its relationship with the goddess or the spiral dance, the cyborg makes no sense. Which of the three has weathered best? The cyborg is being retired, but arguably the goddess and spiral dance can still do work – indeed, Starhawk is hard at work on a campaign to turn *The Fifth Sacred Thing* into a feature film with wide distribution.[4]

And it is hardly irrelevant or accidental that both Starhawk and Haraway have been to Clayoquot, the place where I learned the word "synchronicity." Starhawk came in the summer of 1993 to lead a spiral dance at the camp and was arrested alongside her partner. She has an ongoing history of coming to British Columbia to run annual witchcamps and earth activist training, and many people at Clayoquot had long attended them, so her presence and influence preceded her arrival at the camp.[5] Her novel, *The Fifth Sacred Thing,* was recommended to me by many interviewees, who clearly found it useful in thinking about the events of the summer. Haraway came later, to a workshop of academics and activists, not mutually exclusive categories, that eventually produced the thought-provoking collection *A Political Space: Reading the Global through Clayoquot Sound* (Magnusson and Shaw 2003b).[6] When I returned to Tofino after the workshop had taken place, Valerie Langer of the Friends of Clayoquot Sound mentioned having read Haraway's "A Manifesto for Cyborgs" (Haraway 1985). She appreciated its attention to the lack of purity and the breaching of boundaries, as well as its relevance in a small place such as Tofino, where environmentalists, loggers, and locals all rubbed up against each other and were not always necessarily different people or people with whom one could maintain antagonisms.

Performativity, Genealogy, and Denaturalizing the Nature of Gender

Judith Butler's work on performativity and drag, as well as her account of genealogy, while not explicitly about ecology or environmentalism, or nature as the non-human, offer useful resources, and rather different words to Haraway's cyborgs, dogs, naturecultures, and material-semiotics. *Gender Trouble* questioned the supposed naturalness of being female through an elaboration of the performativity of gender – of the necessary repetitions of gender, which both create a sense of stability and suggest the possibility of refiguring gender. In this way, performativity denaturalizes sex and gender, and in the process, the category of woman. Butler (1992, 16) offered a shift away from essentialisms, strategic essentialism, and anti-essentialisms as ways of defining woman (and nature), "to expand the possibilities of what it means to be a woman and in this sense to condition and enable an enhanced sense of agency." Her work is particularly useful for contesting understandings of eco/feminism as based on a maternalist identity politics and for developing alternative accounts in which motherhood and activism are understood as performative.

Whereas Butler's work on performativity has been much recognized, her articulation of genealogy has received less consideration. Alison Stone (2005, 4) has drawn attention to the importance of genealogy in Butler's performative politics: "Butler understands women to have a genealogy: to be located within a history of overlapping practices and reinterpretations of femininity. This genealogical understanding of femininity allows Butler to propose a coalitional feminist politics, which requires no unity among women but only loosely overlapping connections. For Butler, feminist coalitions should aim to subvert, not consolidate, entrenched norms concerning femininity." Thus, for Butler (1990, viii-ix), genealogy is a critical tool in interrogating the naturalization of being female and in exploring the performative dimensions of gender: "Genealogy investigates the political stakes in designating as an origin and cause those identity categories that are in fact the effects of institutions, practices, discourses with multiple and diffuse points of origin." As Hekman (1996, 4, emphasis in original) summarized, genealogy offers a tool for challenging dualisms:

> That women *are* sex, that they represent the body side of the mind/body dualism, is one of the central tenets of the Western canon. Foucault's approach deconstructs this dualism, redefining sexuality and its truth. Feminists have used this redefinition to question the association of woman with body and sex. Foucauldian-inspired feminists have not, however, tried to move "woman" to the other side of the dichotomy by claiming that women are just as rational as men. Rather they have attempted to subvert the categories that define women's inferiority.

Butler's concerns with naturalization and her performative and genealogical undoing of it are profoundly useful for eco/feminists. Although her work may not seem to be focused on matters of nature, other than woman's nature, and her bodies remain thus far resolutely human, Vikki Bell (2008) underlined the possibilities of her performativity for thinking about ecological survival, through the work of Stengers and Whitehead in particular. Furthermore, to focus solely on the fact that Butler's work does not specifically address ecology is to miss its potential, as I outline next through my genealogy of eco/feminist foremothers, as well as in Chapter 6, which explores the performative drag of the Raging Grannies.

Eco/Feminist Foremothers

The history of eco/feminism is often traced through stories of foremothers – Rachel Carson, author of *Silent Spring*, is one consistently cited example. These are women who would not have used the word "eco/feminism," not least because they lived and worked before the genesis or proliferation of the term. Nonetheless, their lives, work, and practices have been described as eco/feminist.

Patricia Hynes (1985, 1989) linked the lives of Rachel Carson, Lois Gibbs, and Ellen Swallow, describing them as "catalysts of the American environmental movement." Eco/feminist historian Carolyn Merchant (1994) also wrote about Carson, Gibbs, and Swallow, as well as Helen Caldicott, Petra Kelly, Winona LaDuke, Karen Silkwood, and Judy Smith. Literary critic Vera Norwood discussed Carson, the lives of women adventurers, and representations of southwestern US landscapes in women's writing (Norwood 1987; Norwood and Monks 1987). Unsurprisingly, attention to the lives of those who are less known is not so common (Garland 1988; T. Kaplan 1997; Breton 1998; MacGregor 2006).

Hynes's biographical approach to Rachel Carson's life and work raises questions that ought to be of interest to eco/feminists. As Hynes wrote in *The Recurring Silent Spring* (1989, 6), "Until Rachel Carson, historians could still dedicate only a few pages in the annals of the American ecology movement and the history of environmental science to the contributions of women." Reflecting on her efforts to write biographically of Carson, Hynes (ibid., 2) noted that

> the difficulty arose when some of those closest to her work resisted any new biographical work on Carson. They contend she was private, so there is not much to know or tell about her person; that her life was her work, so people should read her work rather than probe her life; that enough has been written about her by those who already knew her and, furthermore, that the definitive work on her already exists, Paul Brooks' *The House of Life.* In addition, some of her personal papers no longer exist; they were destroyed immediately after she died. The letters to Dorothy Freeman, the only other person besides her mother to whom Carson "gave herself completely," as Paul Brooks wrote, are currently unavailable. These letters were written just before and during the years in which Carson researched and wrote *Silent Spring.*[7]

In the biographies of Carson, Hynes (ibid., 3) "found an increasingly stereotyped, one-dimensional, 'unmarried' woman ... and wanted to write a fresh picture of the woman." Hynes's (ibid., 12) own book "debunks the myth that Carson was a woman rich in work, but less than rich – impoverished even – in love. Her work was an expression of her love and she found love where her work was." Hynes paid attention to Carson's close relationships with women – her mother, various science teachers, and Dorothy Freeman, her friend in later life – and with the places she shared with these women. As a child, she learned the names of flowers and birds from her mother as they walked in the woods and wetlands around the Allegheny River in Springdale, Pennsylvania. She got to know Dorothy Freeman as they explored the Maine coast, where she returned each summer. As Hynes asserted, one cannot tell Carson's life story without showing that her love of nature was intertwined with her passionate relationships with mentors and friends.

Rather than asking whether Carson was a feminist or an eco/feminist, Hynes explored her impact, suggesting that having "spawned" a movement might have gratified her. Hynes (ibid., 9) was not afraid to return to fundamental questions about the relationship between feminism and nature:

> With the full benefit of the past two decades of feminism, how would Carson analyze the relationship of men and women to nature? Where would she direct her critique, what connections would she make between pollution and other oppression, most specifically the oppression of women? Or, turning the question of Carson writing today, into one of my reading of her work today, what do we learn from a feminist reading of Carson and *Silent Spring?*

As Hynes (ibid., 4) put it, *Silent Spring* "was activist, not just expository; it was written to reform, not just to have a forum." Her attention to Carson's personal relationship with nature is crucial to understanding the latter's significance for eco/feminism. Hynes's questions about the particularities of Carson's life, her female friends, the places she loved, and about feminism get beyond reductive essentialisms and begin to suggest the potential of understanding eco/feminist matrilineages as a genealogical practice. Hynes's work also points to the usefulness of a biographical approach for the making of eco/feminist theory. This is despite criticisms by some eco/feminists, who bemoan the "ritual citation of Rachel Carson as the foremother of motherhood environmentalism" (Sandilands 1999, xiv). For Sandilands (ibid., 212n8), "Carson was a scientist and author; ironically, she spent her life trying to overcome the assertions of feminine particularity that motherhood environmentalism heaped upon her. Carson may have cared deeply about human and non-human life, but it was not a care reducible to household activities." This curious passage, which insists on defining Carson as "scientist and author," excises her life work as caregiver for her mother, her nieces, and her grand-nephew, and positions her as passively accepting an unwanted role as carer for various relatives – rather than recognizing how the tension between caring and "working" affected her life. Patricia Hynes (1985, 1989), Vera Norwood (1987), Linda Lear (1997), and Nancy Unger (2004) offer rather different ways of looking at Carson's life and work. Unger (ibid., 54-55) demonstrated the heterosexism of responses to *Silent Spring,* which stressed Carson's gender, sexuality, and "unnatural" childless status. Because she was not a mother, she was not allowed to speak for nature – one adviser to Eisenhower even wondered "why a spinster with no children was so worried about genetics" (ibid.). Norwood, in particular, provides a nuanced reading of "home" in Carson's work, noting that she constantly redefined nature and home. Hynes's work consistently situates Carson in a complex web of relationships, which included biological family as well as female friends and mentors and intense relationships with place. My point here, in paying attention to Carson's life, is to link the practice of turning to eco/feminist foremothers with feminist practices of generating matrilineages and in this process to suggest the radical potentials of such practices.

Matrilineages as Feminist Genealogies

Virginia Woolf offered a classic statement of feminist matrilineage in *A Room of One's Own* (2002, 69), her meditation on the challenges facing women writers.[8] She proposed that she and other women writers "think back through our mothers." On taking a book from a shelf in the British Library, Woolf (ibid., 79-80) noticed that

> it stood at the very end of the shelf, was called *Life's Adventure,* or some such title, by Mary Carmichael, and was published in this very month of October. It seems to be her first book, I said to myself, but one must read it as if it were the last volume in a fairly long series, continuing all those other books that I have been glancing at ... And I must also consider her – this unknown woman – as the descendent of all those other women whose circumstances I have been glancing at and see what she inherits of their characteristics and restrictions.

Woolf pointed to the impossibility of reading books by women in isolation, suggesting that earlier texts were implicated in the emergence of later ones and that recent texts offered a contemporary context through which earlier texts could be read as constituting a history of women's writing.

Matrilineages such as Woolf's might imply the creation of a feminist canon consisting of women writers whose work can be found in the British Library. Yet Woolf clearly had some understanding of exclusions, as suggested by her fictionalized reflections in *A Room of One's Own,* on the challenges of being a woman writer, and the difficulties of obtaining an education, including her father's refusal to support her formal education. Nonetheless, Alice Walker's essay "In Search of Our Mothers' Gardens" (1983) can be understood as both in continuity with and an engaged critique of Woolf. Like Woolf, Walker complicated notions of who counted as a literary foremother by extending the meaning of "mother" beyond her own mother to other female relatives and friends and neighbours, and to significant women who had inspired her. For both Woolf and Walker, the process of thinking back through their mothers was rather an expansive gesture toward a wider women's history. For Walker, this included documenting black women's creative practices. Her foremothers differed from those of Woolf, and their work was not to be found on the shelves of the British Library. Walker (ibid., 235) quoted Woolf but added her own critical commentary in square brackets:

> For it needs little skill and psychology to be sure that a highly gifted girl who has tried to use her gift for poetry would have been so thwarted and hindered by contrary instincts [add "chains, guns, the lash, the ownership of one's body by someone else, submission to an alien religion"], that she must have lost her health and sanity to a certainty.[9]

Walker extended what counts as art and creativity beyond the strictly literary, or written, reading her own mother's gardening as an expression of creativity, and an important one at that, given the injunctions against writing during slavery. Whereas Woolf searched the shelves of the British Library for women writers and did eventually find some, Walker had a different task. Her stories counterposed her research into the life and work of white Southern writer Flannery O'Connor with her search for her own mother's home and for traces of black women writers Phyllis Wheatley and Zora Neal Hurston. Their stories were not to be found in any library; following a hand-drawn map and verbal directions, Walker had to trace a neglected route to Hurston's grave, by then covered in weeds.

The project of constructing eco/feminist genealogies points to a matrilineal epistemology that is more complicated than critics sometimes allow. For instance, in "Happy Families? Feminist Reproduction and Matrilineal Thought," Linda Williams (1992, 44, 45) voiced concerns about feminist, or female, matrilineages (and notions of sisterhood) as "both a separatist ideal and an organizing metaphor for women's studies," and what she saw as the persistence and attraction of the notion of "unmediated feminine communication":

> Why do we so employ familial metaphors to interpret our conceptual and scholarly relationships with each other? What are the power relations at

> stake in setting up feminist networks of thinking which rely on mother-daughter or sisterly ties? Why are we so reluctant to rid ourselves of the family? These questions focus not only on the problem of mother-daughter relations in history or psychoanalysis, but crucially on the way we have interpreted women's literary history as a family history, glued together by those "unknowable" feminine relations ... Thus it seems, ironically, that the very force with which some writers have drawn upon to signal the breakdown of patriarchal family relations – a feminine communication which disrupts normal epistemologies – has then been used to make coherent an alternative Great (female) Tradition.

In referring to Woolf and Walker, Williams (ibid., 51) was particularly critical of what she read as their "ideal but exclusive form of intra-subjective psychic contact as a model for literary history (or scientific knowledge) – the notion that down the historical chain of women's writing an ideal form of female creativity is communicated and regenerated." However, her reservations are not borne out by the texts themselves. Williams's (ibid., 50) particular citations – Woolf's "we think back through our mothers if we are women," and Walker's "our mothers and grandmothers have, more often than not anonymously, handed on the creative spark, the seed of a flower they themselves never hoped to see: or like a sealed letter they could not plainly read" – both highlight the *work* involved in interpreting and imagining the creative lives of their foremothers. Walker's research on Zora Neal Hurston demonstrated both the necessity of creative labour in generating documents of her life and the challenges of imagining lives that are not recorded in formal histories. Likewise, even Woolf's task of finding women authors in the British Library points to the research and work involved in "thinking back through our mothers." Furthermore, we might now apprehend that the family is rather difficult to jettison and focus instead on critiquing its hetero-patriarchal nuclear version.

Unlike Williams, Laurie McMillan does not require one to reject matrilineages or to accept them as naturalized or naturalizing. She (2004, 113) understood Walker as "signifying" on Woolf, meaning "a critical approach with African American roots that enacts repetition with a difference. The repetition pays tribute to the precursor writers, while the difference is a way of revising the precursor's story or practice, often exposing its limitations." McMillan (ibid., 112) read Walker's work as *performative* rather than naturalized:

> Naturalized personal criticism would use an autobiographical voice that presents itself as transparent, as if reading about a person is to fully know and understand that person and his or her experiences. "Performative" personal criticism, on the other hand, highlights the way an identity is taken up and used in a certain way, drawing attention to autobiography's mediation through language and cultural context. As a performative approach keeps people from being reduced to their representations, notions of identity remain fluid and changeable; an essentialist identity politics can then be resisted rather than enacted through the autobiographical criticism.

For McMillan (ibid., 113), the use of personal experience may be either revolutionary or conservative, but echoing Spivak and Sawicki, she argued that essentialism is not a matter of prior theoretical judgment. When this citational practice is read as both homage and critique, as an insistence on connection and a careful marking of difference, understanding the construction of feminist matrilineages becomes more challenging. Walker's reading of Woolf reveals the practice of matrilineages not just as a challenge to the patriarchal heterosexual nuclear family, but also as gesturing to the fractured bonds between women. Far from being romanticized connections, matrilineages are the fraught construction of relations between women. Thus, feminist matrilineal practices must also be understood as intentionally generative, as involving the creating and imagining of (non-innocent) feminist community.

Woolf's and Walker's accounts of their work can be understood both as instances and critiques of Foucault's (1984, 76) description of genealogy as "grey, meticulous and patiently documentary. It operates on a field of entangled and confused parchments, on documents that have been scratched over and recopied many times." Walker's search for traces of Zora Neal Hurston, in stories, shreds of conversation, and half memories that led to her grave and the site of her house, acts as pointed commentary on the very absence of parchments and documentary evidence in many lives; the colours of her mother's garden stand in stark contrast with Foucault's grey documents.

Feminist matrilineages might best be understood, not as reproducing a conservative hetero-nuclear family, but rather as feminist genealogies in the making, bringing to attention the creation of relationships between women. A genealogical reading makes explicit the possibilities of constructing and reading matrilineages as not always and inherently essentialist.

Woolf and Walker were crafting feminist knowledge practices of a kind echoed by Adrienne Rich's attention to re-visioning.

What becomes apparent in the *work* of Woolf, Walker, Rich, and others is that whereas Foucault's genealogy is a critical historiographical practice that understands the researcher or writer of history as always implicated in its production, feminists have pursued this approach to its logical conclusion, actively engaging in history making and history writing to imagine and create shared community. Teresa de Lauretis (1993, 402) pointed to this ongoing process, reflecting that her own work as a woman scholar was embedded in

> a scattered, fragmented and yet historically embodied lineage of female thought and writing which my sometime compatriots, the Italian philosophers of the Milan Women's Bookstore, have called a genealogy of women, and others call by the name of feminism. That is neither a tradition nor a kinship of dispossessed mothers and daughters, but rather the trace of a discontinuous, elusive and daily reconstructed feminist genealogy.

As de Lauretis (1994, 13) explained, feminist genealogy is centrally concerned with formulating "a freedom for women that ... is generated, indeed engendered, by taking up a position in a symbolic community, a 'genealogy of women,' that is at once discovered, invented, and constructed through feminist practices of reference and address," again rebutting easy recourse to notions of unmediated communication through history. I link this with the take-up of genealogy by various feminists (de Lauretis ibid.; V. Bell 1999; Stone 2004; Hemmings 2005, 2011) as a way of critically conceiving of feminism, its histories, and its reworkings of these histories as the intentional construction of an "imagined community," of a "re-visioned" community of feminists. Crucially, this does not rely on essentialist or universalist pretensions. Therefore, matrilineages can be approached, not as naturalized accounts of familial or reproductive relationships between women, but rather as stories that bring feminist community into being. This is in keeping with Alison Stone's (2004, 137) reworking of genealogy in Butler's performative politics:

> I will suggest that women always become women by reworking pre-established cultural interpretations of femininity, so that they become located – together with all other women – within a history of overlapping chains of interpretation. Although women do not share any common

> understanding or experience of femininity, they nevertheless belong to a distinctive social group in virtue of being situated within this complex history. This rethinking of women as having a genealogy entails a concomitant rethinking of feminist politics as coalitional rather than unified. According to this rethinking, collective feminist activities need not be predicated on any shared set of feminine concerns; rather, they may arise from overlaps and indirect connections between women's diverse historical and cultural situations. I hope that my exploration will begin to show how a genealogical rethinking of women could enable feminists to oppose (descriptive) essentialism while retaining belief in women as a group with a distinctive, and distinctively oppressive, history – an ongoing history which is an appropriate target of social critique and political transformation.

Stone (ibid., 147, emphasis in original) noted that Butler and Moira Gatens, whom she also cited,

> appear concerned to trace the history of the concept "woman" rather than the history of women themselves. Yet the concept of genealogy can only provide a way of grasping women as a distinct (albeit non-unified) social group if women *themselves* have a genealogy – that is, to anticipate, if their experiences and psychologies are shaped in overlapping and historically interconnected ways. Thus, any anti-essentialist appropriation of the concept of genealogy must clarify the relationship between the genealogy of ideas concerning femininity and the genealogy of women themselves.

So, though much of this chapter has focused on genealogies of eco/feminism, I also want to follow Stone's lead and pay attention to genealogies of *eco/feminists.* I have already discussed Carson, and I continue this in my attention to the stories of women activists at Clayoquot. As Stone makes clear, these are not to be read as individualized autobiographies of a neo-liberal subject. In approaching women's oral histories genealogically, then, I understand eco/feminist matrilineal practices not as implying the "canonization" of any particular individual such as Carson, who would surely have rejected it, but rather as a way of exploring connections. This approach echoes Kamala Visweswaran's (1994, 8) argument that "linking the individual to the universal (especially when the notion of the universal has all but collapsed) is no longer adequate, and in situating an individual within a particular community, the local and specific are broached in ways we might well term ethnographic. Caren Kaplan thus reads ethnography as a means of challenging the genre-structure of autobiography."

Experience as Essential/Essentialist

Such a project also requires an exploration of the status of experience in feminist theorizing. Like the categories of gender, woman, women, and nature, invocations of experience and especially experience of, or in, nature are commonly deemed essentialist.

Experience has been central to feminist theory and politics. Indeed, Teresa de Lauretis dated the advent of feminism to women's sharing of their experiences: "Experience is the basis of feminism in the sense that feminism began the moment women started talking to each other" (quoted in Skeggs 1995, 15). However, basing feminist theory in women's experiences was ultimately seen as intensely problematic, and feminists challenged notions of the authority of experience as evidence and epistemology (Mohanty 1992; J. Scott 1992). Often the use of experience in feminist theory was taken to imply essentialist notions of truth, identity, and self, to suggest an uncritical equation of experience with authenticity. As a result, experience and essentialism have been linked and critiqued as a foundational politics, rendering invocations of experience particularly fraught for eco/feminism.

Joan Scott's essay "Experience" (1992) is a classic reference here. Scott, who played a key role in refiguring understandings of experience and recuperating it for feminist analysis, noted that it was used to naturalize or reify difference, especially differences that mattered to feminists, such as gender, sexuality, and race. For Scott (ibid., 25), such histories of difference "take as self-evident the identities of those whose experience is being documented and thus naturalize their difference." Scott (ibid., 25-26) pointed to the need to focus on processes of subject creation, not just experience itself, and on the researcher as well, the historian in her case, who is doing the interpreting: "It is not individuals who have experience, but subjects who are constituted through experience. Experience in this definition then becomes not the origin of our explanation, not the authoritative (because seen or felt) evidence that grounds what is known, but rather that which we seek to explain, that about which knowledge is produced." For Scott (ibid., 37), "experience is always already an interpretation and in need of interpretation." Some have taken her essay to suggest that women's experiences can no longer provide a foundation for feminist theory. Indeed, the notion of unmediated experience as foundational has been challenged, but I understand Scott's work as contributing to a more careful approach to women's experiences. It is worth bearing in mind that her essay appeared in a collection titled *Feminists Theorize the Political* (Butler and Scott 1992). Beverley Skeggs (1997, 28) also argued that such a reformulation of the meaning and status of experience enables a shift from "experience as a foundation for knowledge to experience as productive of a knowing subject," in which identities are continually in production rather than being fixed.

In what reads as a diagnosis of eco/feminism's ills, Stephenson and Papadopoulos (2006, 20) starkly warned of the consequences of ignoring experience: "Ultimately the cost of jettisoning a close examination of the complexity of experience is to deter research which can engage with and contribute to emerging modes of political engagement." Here we have some indication of the perceived problems with eco/feminism; the focus on essentialism forecloses a serious engagement with emergent politics. Given the association of essentialism and women's experiences, it is not surprising that the literature on eco/feminism has paid little attention to empirical research, though there are some notable exceptions, one of which is Noël Sturgeon's *Ecofeminist Natures* (1997). Sturgeon drew on her research in activist settings to offer an account of the complexity of eco/feminist activism and to produce evidence that eco/feminism was not reducible to essentialism. Sturgeon's work demonstrates that "it is both practical and political to think about and carry out activism without the belaboured dichotomy between theory and practice," as Katie King proclaimed on the cover of *Ecofeminist Natures* (ibid.).

Other instances of empirical research exist, some of which invoke notions of women's experiences as a source for eco/feminist theory, and this is what interests me here. Greta Gaard's *Ecological Politics: Ecofeminists and the Greens* (1998) focused on the US Green Party and explored what it terms "ecofeminist roots" – women's paths to eco/feminism, including peace and anti-nuclear activism, feminist spirituality, animal liberation and environmental campaigns, the anti-toxics movement, and finally, "the path of lived experience." Gaard (ibid., 29) acknowledged that "of course, women who may or may not have been activists began making the connections to ecofeminism from their own personal experiences." However, she (ibid., 30-31) did not elaborate on the importance of experience, writing that "the paths of lived experience are not drawn on ... simply because they are too numerous; there are different paths for every person." Gaard (ibid., 33) named only three brief instances of "lived experience" that she encountered in the eco/feminist literature: Charlene Spretnak on the "post-orgasmic state," Karen Warren on her rock-climbing experiences, and Lourdes Arguëlles on her sense of childhood connections with the animal world. The scarcity and idiosyncrasy of this list highlights the curious absence of attention to women's experiences in many eco/feminist texts.

Sherilyn MacGregor (2006, 129), one of the few eco/feminist academics to address experience and to highlight methodological issues in relation to eco/feminism, identified that "a methodological issue that is of great significance to ecofeminist research on women's activism is the interpretation and presentation of 'lived experience.'" As Macgregor (ibid., 236) noted, her decision to theorize the lived experience of what she termed "quality of life" activists and advocates in Toronto was unusual: "Inviting an encounter between an expression of ecofeminist ideas about women activists and women activists themselves is something that, to my knowledge, no other ecofeminist scholar has done."

Some earlier eco/feminist scholarship is empirical, if not always explicitly so. Carol Adams's (1991, 1994a, 1994b) focus on women and animals was informed by her work with victims of domestic violence and her involvement in the refuge movement. In *Neither Man nor Beast,* Adams (ibid.) wrote of how crucial pets – companion animals – could be to women who were isolated through domestic violence, and of how this relationship could be exploited and abused by violent men, who threatened, tortured, and killed pets as part of their violence to women. The work of Adams is an important, and rare, instance in which eco/feminist theory emerges from an exploration of particular women and their everyday relationships, here with animals. Although Adams (1991, 141, n.2) does not extensively focus on the methodological aspects of her work, she did stress "the importance of first person narrative in (eco)feminist theory building."

Other work that offers useful resources in refuting critiques of essentialism has focused on (non-human) nature, history, narrative, and genealogy. These include Stephanie Lahar (1991), Terry Tempest Williams (1991), Carol Bigwood (1993), Linda Vance (1993), Jim Cheney (1994), and Carolyn Merchant (1996). All have addressed how we might come to know more about "nature" and women's relationships with nature through how we might tell better stories. Rather than attempting to disprove essentialism, these works focus on exploring – and thus expanding possibilities of – relationships with nature. Some of them can be seen as precursors to research that now gathers under various rubrics, one of which is the "posthumanities" (for instance, see Haraway 2008, 74, 332-33).

Eco/feminism's lack of attention to empirical research, and specifically to women's experiences, is disappointing, given the emphasis on autobiography, biography, and stories of women's lives in feminist work more generally, and indeed, given the attention to eco/feminist foremothers. Sturgeon (1997) and Macgregor (2006) reflect on the empirical in eco/feminism in differing ways, and both attribute its absence to a tension between theory and activism. Sturgeon's direct theory challenged any notion that activists do not also theorize. MacGregor expressed concern over how "lived experience" has been theorized in feminism more generally and worried that women's experiences have been taken uncritically as a basis for eco/feminist theory. Her aim was to do more than just celebrate the "triumphs" of activists and to show the difficulty of engaging in activism while running a household. Taking seriously MacGregor's reservations about an uncritical take on women's experiences, I turn to elaborate feminist thinking and rethinkings of experience.

Eco/Feminist Oral History: From the Question of Nature to the Nature of Questions

Focusing on the question of essentialism may be useful for understanding the fractured relationship between eco/feminism and feminism. However, I suggest that it is much less useful when applied to women's environmental activism or their relationships with nature. Considerable confusion has arisen over this point, and part of my intention is to clarify this matter. So my interest in trying to articulate what eco/feminism might be centres not on defining or typologizing theories of eco/feminism, or on assigning women or their utterances to the categories of essentialist, anti-essentialist, or strategically essentialist. Rather, I am interested in stories, in histories, in narratives, and specifically in the oral histories of those at Clayoquot.

At the same time, the complexity of theorizing experience has led some researchers to shy away from using accounts of women's experiences, such as oral history. Cultural theorist Ann Cvetkovich provided some insightful reflections on her coming to oral history and is one of the few who explicitly reveal their ambivalence regarding experience and oral history. In her *An Archive of Feelings: Trauma, Sexuality and Lesbian Public Cultures* (2003, 165), she reflected: "I came to oral history with a certain amount of resistance given that my theoretical background had taught me to be suspicious of what Joan Scott calls 'the evidence of experience.'" A key issue for

Cvetkovich was a critique of the idea that the disempowered have any privileged access to explaining the conditions of exploitation. Informed by her reading of key feminist texts such as Gayatri Chakravorty Spivak's "Can the Subaltern Speak?" Cvetkovich (ibid., 166) also drew on the argument that "one of the great, and often misunderstood lessons of deconstruction is that far from undermining the grounds for inquiry, it is at its most interesting when applied to concrete decisions such as those demanded by doing oral history." In this, she refers to a vital dimension of oral history, one of its most distinguishing and generative features.

Cvetkovich's admission of ambivalence regarding oral history is suggestive of others' aversion to it. Nonetheless, she remained open to its possibilities. Musing that she was more accustomed to using archives than to creating them, she (ibid.) also reflected that

> mixed in with my scepticism about oral history were curiosity and fascination. I was driven by the compulsion to document that is so frequently engendered by the ephemerality of queer communities and counterpublics; alongside the fierce conviction of how meaningful and palpable these alternative life worlds can be lies the fear that they will remain invisible or lost. Oral history can capture something of the lived experience of participating in a counterpublic, offering, if nothing else, testimony to the fact that it existed. Often as ephemeral as the very cultures it seeks to document (since both tapes and transcripts are records of a live event that is past) oral history is loaded with emotional urgency and need.

For me, oral history offered the opportunity to document a history, or herstory, of Clayoquot, to bring into being the stories of women, and of eco/feminism, that might not otherwise be told. I was drawn to feminist oral history's doubled approach to the *making* of history – that it documents the stories of women and is informed by a strong understanding that unless feminists write and make this history, it will not exist, the stories will not be told, this history will not enact the world. As Niamh Stephenson and Dimitris Papadopoulos (2006, xvi) noted, "If we accept that social research is not simply a matter of representation, but involves actively intervening in and constructing our current social and political conditions, we are then presented with a series of problems about how this happens in the process of researching experience." Likewise, Joan Scott's orientation to critical research (with its emphasis on process, interpretation, and contingency) advocates a constitutive understanding of experience and subjectivity. Her move toward Foucault in "Experience" (1992) situates the use of experience firmly within the logic of genealogy rather than as a foundation for politics. That is, the focus shifts from "this is my experience, and therefore it's not anyone else's to critique" to "this is my experience, and in the telling, I show how social and political forces are at play with how, why, and to what end I tell it." In this way, experience is called into question rather than taken for granted.

Narratives have long been of interest in accessing an individual's subjectivity, experience, and reflections on the past. They have played a central role in the development of oral history as a practice and methodology. Narratives foreground the emergence of the subject, as Byrne (2003, 30) succinctly summarized:

> An interesting conjunction has emerged in theoretical and particularly in feminist scholarship between a concern with subjectivity and the self and with narratives. The interest in narratives and the narration of identity signifies a move away from the search for essential, universal or even rational identities and a stress on the more uncertain and creative processes of construction and fabrication.

Byrne (ibid.) added that

> approaching processes of subject construction through narrative analysis has the advantage of acknowledging the constructed, flexible and fictionalized nature of the process of accounting for the self. The subject is understood as in process. Narrative approaches also enable the exploration of processes of subjection – the ways in which certain subject positions (ways of being and ways of being recognized or acknowledged by others) – are discursively available for individuals to occupy. The fiction of a whole or coherent self may be created in the process of narrating the self, yet it can also be undermined in the telling.

Thus, narrative offers a promising alternative to categorizing accounts according to whether they are essentialist, anti-essentialist, or strategically essentialist.

It should be clear that my intent was not to "give voice" to the Clayoquot activists, not least because this was unnecessary – many of them have written their own stories. My intention was first and foremost to listen and to resist the imperative to collapse what was said into essentialism and maternalism. As will become clear throughout this book, I treat women's accounts of themselves as genealogies, as undoing any straightforward telling of their lives and thus as genealogies already, though in need of further genealogical elaboration, to paraphrase Joan Scott. In this I echo Maria Tamboukou's explorations of how "women's autobiographical writings can inform the writing of a feminist genealogy" (Tamboukou 2003, 5). Thus, I treat interviews much like academic texts, where, of course, I do not need to give voice to theorists. Rather, following Katie King (2010), "More than most essays, this one relies upon many quotations. The point is to include other worlds in their own words in our game of cat's cradle." I engage in conversations with theorists, some of whom are activists and do not write academic texts, as well as with many academics, some of whom would also consider themselves activists.

Oral histories are often collected for others. The urge to archive is motivated by my sense that these stories are not mine, not for me to own and keep. The impulse to archive is also motivated by the sometimes desperately obvious realization that I cannot possibly do justice to all the stories or even to/for any one of them. I understand the process of doing justice to the stories as a collective activity. I cannot do it on my own. Minimally, it involves the reader and other researchers. In her account of feminist archiving, Kate Eichhorn (2010, 641) stated that "to label something an archive remains a powerful authorizing act," stressing its capacity as "a structure that determines which statements can and do act in and upon the social world."

I assemble the narratives to achieve a particular end – to tell a story of Clayoquot and of eco/feminism, with the necessary recognition that it is hardly the only one that can be told. As Visweswaran (1994, 8) has reflected on ethnography, situating an individual within a community can challenge the genre of autobiography, and I suggest that the same is true for this kind of collective oral history.

Relatedly, though some might take oral history to imply a humanist voice, an individualist narrative, my turn to experience and oral history is not to demonstrate people's capacity for acting heroically. Here I echo Stephenson and Papadopoulos (2006, 136) in their critique of "illness narratives," which they noted "are typically valued for the extent to which they can give meaning to lives and open possibilities for people to act agentially, to increase their capacity for action and take some measure of control over their lives." Like Stephenson and Papadopoulos, I resist heroic narratives – indeed, a few interviewees made this point themselves, in indicating that they would willingly speak about the campaign, but were less keen on my interest in biography and life story. Initially, I puzzled over this distinction, for it seemed to me that my approach was precisely trying to refuse any notion of individual human(ist) agency or heroism, though I was not necessarily able to articulate this well at the time.

From "Women and Nature" to "Our Relationship with Ourselves"

> she keeps insisting herself on the telling
> because she was telling me right from the
> beginning stories out of a life are stories,
> true, true stories and real at once – this is
> not a roman/ce, it doesn't deal with heroes
>
> – DAPHNE MARLATT, *ANA HISTORIC* (1990)

Shane Phelan (1993) proposed the concept of "intimate distance" for refiguring our understandings of our relationships with nature. In reminding us that it is necessary "to ask the nature of nature," she pointed out that environmentalists – and, I would add, feminists and eco/feminists – commonly speak as if they know what nature is and that this nature has often ending up meaning non-human nature, even for those who wish to include humans in their understanding of nature; nature remains "other." Phelan (ibid., 58) suggested that we address this challenge "by foregrounding the construction of the ecological community as a human activity, and by seeing humans not as 'members' or as 'different from' nature, but as elements of a dynamic and never fully locatable process." In viewing nature as a category that "we can deconstruct, but not eliminate," she (ibid., 45, 57) also stated that "basing our arguments upon nature is dangerous, unless nature is reformulated." For Phelan (ibid., 58-59), this approach suggested that "the questions to be asked then are not, What should humans do (or not do) to nature? but instead, How do we understand ourselves and our world? How ought we to negotiate our relationships with ourselves?" The implications of this are echoed in Éric Darier's (1999, 27) account, drawing on Foucault's work for thinking about ecological ethics and subjectivities:

> Foucault's non-ethics has important consequences for environmental ethics in that the focus shifts away from the presumed discovery or "rediscovery" of a true permanent "ecological self" to the active constitution of

> subjectivities which constantly rework humans' relations with themselves, with other life-forms, and with the world generally. The ethical constitution of what might be called "green subjectivities" might be the endless process of "ethicization" of being human in the world.

As Irene Diamond and Lee Quinby (1988, xiv) elucidate, "Such an ethic supplants human-nature arguments with genealogical investigation." Genealogical critique offers a methodology that refuses a priori theoretical pronouncements about essentialism, as well as questions about relationships between women and nature, positing such approaches to be founded in dualisms. Instead, it foregrounds methodological and empirical investigation. Thus, the future of eco/feminism, feminism, and the planet might not be secured through theoretical pronouncements of, and insistence on, anti-essentialism (and the exclusion of offending varieties) but might better be secured through genealogical investigation.

Whereas Haraway turns specifically to the multiple stories of science, science fiction, and women primatologists, ultimately I turn to activists gathered at an eco/feminist peace camp on the west coast of Canada in the summer of 1993, many of whom had been to Starhawk's witchcamps and who joined her spiral dance. I examine apparently overdetermined figures – mother, child, nation, home, family – and have no cyborg or dog in my back pocket as alternatives. I bring together the spirit of the cyborg, of Haraway's naturecultures, and Butler's performativity and genealogical practice with Starhawk's immanent goddess to think through the apparently more domestic figure of women environmental activists. The question then is not what women should or should not do to nature, what feminism or feminists should do to nature, or whether women are closer than men to nature. I am reminded of Barbara Gates's (1998, 20) reflection that "inherent in ecofeminism is a belief in the interconnection of all living things. Since all life is nature, no part of it can be closer than another to 'nature.'" Thus, the relevant question might be how we understand ourselves and the world. And so, I suggest that oral history is a valuable approach to exploring eco/feminist subjectivities, which might also be more-than-human(ist) subjectivities. I did not, therefore, ask women whether they were closer than men to nature or more concerned about the environment. Nor did I ask them to speak about their relationship with nature, trees, forests, and wilderness, or to discuss whether maternalism spurred them to activism, whether they were concerned about the future of their children or grandchildren, whether their actions were essentialist, or whether they were anti-essentialists or cyborgs or queers or citizens, as some have suggested. Neither does this book focus solely on the blockades, arrests, and incarcerations, or even the Clayoquot Peace Camp itself. Though some may perceive the blockades and court proceedings as the key site of politics here, I am not so sure. To centre these aspects of Clayoquot would be to circumscribe the site of, and possibilities of, the political. Rather, I asked those whom I met to tell me the story of their lives, including how they came to be in Clayoquot – this is a book about women and their relationship with themselves, selves that are not all human(ist).

4

Clayoquot Histories

Our Home and Native Land?

We came, we sawed, we conquered.

– POSTER OUTSIDE A MacMILLAN BLOEDEL OFFICE (1993)

SUPERNATURAL B.C.

Supernatural B.C.
Where the forest used to be
If you want to see our ghosts
Come and visit our west coast
Supernatural B.C.
Where the fish are history
If you want to see a whale
You'll find three in Sealand jail.

Supernatural B.C.
It's a land of fantasy
We'll have Disneyland instead
When the natural world's all dead
Will it no' come back again?
Will it no' come back again?

Supernatural B.C.
Where the forests used to be.

– JEAN McLAREN AND HEIDI BROWN,
THE RAGING GRANNIES SONGBOOK (1993)[1]

This chapter does not aim to offer a history of Clayoquot Sound, background to the campaign, or an account of the events that led to the blockades. Rather, I focus on how the history of Clayoquot has been told and contested by different parties, at different times, to different ends, and how and why these contested histories matter. In particular, I pay attention to the Friends of Clayoquot Sound (FOCS), arguing that its campaigning against clear-cut logging was intensely and intentionally, as well as inevitably and unavoidably, involved in challenging dominant histories of Clayoquot – and Canada. My point is that history was not what came before the campaign, but that the history – and future – of Clayoquot was precisely one of the sites of contestation in the campaign. The campaign made – and remade – history in more ways than one, challenging dominant narratives and engaging with other, alternative, histories of the region, in order to imagine alternative futures, for British Columbia and for the world.

Logging company MacMillan Bloedel offered its version of BC history in a poster outside one of its offices. Depicting the felling of a giant cedar, the poster read "we came, we sawed, we conquered" (Vidal 1993, 24). In this way, the company echoed the dominant history of Clayoquot Sound, which, like much of the history of Canada, is typically told as one of settlement, development, and progress, of the inhabiting of a barren landscape by the bearers of civilization. Conventional histories of Vancouver Island differ little in narrative form from the conventional history of the world, which tells of the triumph of development and the march of progress. According to that version of history, in Clayoquot Sound environmentalists are preventing the logical outcome of development in British Columbia of increased wealth and prosperity through rising employment – forestry is the history and future of the province.

Yet, I argue that understanding the work of the FOCS as a campaign to preserve rainforest, or wilderness, would be to offer a very limited account of what was at stake; at the very least, the campaign has been about a specific logging practice – clear-cut logging. Although a few early newsletters from the FOCS suggest there were discussions about calling for Clayoquot to be granted national park status, this direction did not take hold, particularly with the realization that park status did not necessarily protect forests and did not actually preclude clear-cut logging – the nearby Pacific Rim National Park was logged before, as well as after, being declared a park. Furthermore, there was an appreciation that national parks were predicated on the concept of Crown land and that much of the land in BC is Crown land, meaning that the provincial government does not even have the kinds of intensely contested legal treaties that exist for other parts of Canada.

The forest industry and the governments that rely on it are critical of environmentalists' apparent penchant for wilderness because they see this as disrupting economic growth, but they are not environmentalists' only critics. Environmental historian William Cronon (1996, 7) has suggested that environmentalists' struggles involve "getting back to the wrong nature." Writing that the "time has come to rethink wilderness," he acknowledged that his suggestion would seem

> a heretical claim to many environmentalists, since the idea of wilderness has for decades been a fundamental tenet – indeed, a passion – of the environmental movement, especially in the United States. For many Americans wilderness stands as the last remaining place where civilization, that all too human disease, has not fully infected the earth. It is an island in the polluted sea of urban-industrial modernity, the one place we can turn for escape from our own too-muchness. Seen in this way, wilderness presents itself as the best antidote to our human selves, a refuge we must somehow recover if we hope to save the planet.

Cronon was writing of a specifically American wilderness and US environmentalism, whereas Atwood was discussing the Canadian wilderness (see Chapter 1); nonetheless, the question of whether environmentalists are getting back to the wrong nature might still be asked of activists in Clayoquot. Yet I am arguing that this kind of account does some injustice to the cultural work in which environmental activists are involved and to the extent to which they might recognize, and fully understand, themselves to be involved in this work. Although Cronon surely had a point, it is also the case that environmental historians have tended to provide accounts of landscapes of the past and how they came to be that way, and have focused less on the historical and cultural work in which environmental activists are engaged (Cohen 1996; Hays 1996; Lipschutz 2001). Similarly, despite, or perhaps because of, the particular version of the "social" invoked, social movement theorists have tended to emphasize the political and legal work in which activists are engaged. Neither approach fully attends to the role of environmental movements in challenging the culture of nature, or naturecultures (after Haraway 2008), of everyday life.

Although Cronon's (1992, 1349) account of environmentalism seems limited, his thoughts on the then emerging discipline of environmental history are useful here (and prefigure more recent attention to the more-than-human):

> On the one hand, a fundamental premise of my field is that human acts occur within a network of relationships, processes and systems that are as ecological as they are cultural. To such basic historical categories as gender, class, and race, environmental historians would add a theoretical vocabulary in which plants, animals, soils, climates, and other non-human entities become the coactors and codeterminants of a history not just of people but of the earth itself.

Environmentalism might usefully be understood as a social movement that also takes seriously the movements of the land. This suggests the redefinition of the category of environmental history to include the *work* of the FOCS and an expansion of the understanding of environmental activism more generally.

Although I foreground the work of the FOCS in this book, there were other environmental organizations involved at Clayoquot, including the Wilderness Committee (then the Western Canada Wilderness Committee), Greenpeace, and Forest Action Network. Here I would also include the Women's Environmental Network in the United Kingdom, which at the time was unusual among environmental groups in leading a campaign against logging of temperate rainforest, when many organizations were focused on tropical rainforest. In pointing to the range of groups involved, I do not want to suggest that the FOCS was necessarily somehow more (or less) radical than the others or to set these organizations against each other. That too would be to mistake some of what went on, not least the intense shared networks of the women in these groups and the ways in which they supported and mentored each other. Rather, what becomes relevant is the extent to which local activists "attempted to break out of this frame and repose the issues inclusively" (Magnusson and Shaw 2003, 271). As Magnusson and Shaw clearly recognized, what makes Clayoquot so interesting is that there was so much more going on than a repetition of a desire for wilderness. My intention is to trace how the Friends of Clayoquot Sound located its campaigning in the context of a range of other histories and thus to foreground aspects of its work, which remain out of sight if one pursues wilderness imaginaries or focuses too intently on the mediated spectacle of the blockades.

I argue that, against the progress narratives of the forestry industry, the FOCS challenged conventional histories of settlement, development, and growth by pointing to the degradation of local landscapes, the devastating effects of deforestation on local communities, and to a local and provincial government deeply entangled with the logging industry. At the same time, the FOCS pointed to Clayoquot's, and Canada's, implication in various global narratives about deforestation. In this way, it provided a profound challenge to the metanarratives of colonial histories of Clayoquot and revealed the official histories of British Columbia to be a very particular account of the region. The campaigning of the FOCS was intensely involved in reshaping the nexus of nature/wilderness/nation through its activism. In this way, the work of the FOCS can be linked with a host of other efforts to challenge the dominant histories of British Columbia – from First Nations, from workers and labour activists, from feminist historians, novelists, and artists, and from the land itself – which all counter any triumphant progress narrative. The FOCS was keenly aware that "throughout Canada's history the economy was based on penetrating resources, appropriating Native lands, and importing labour for the purpose of nation-building" (Mackey 2000, 125). The Friends of Clayoquot Sound, in its challenges to official histories, revealed "tactics of resistance" (Darier 1996) to these dominant narratives and demonstrate the possibilities for ordinary people to make history, and in this process to envision different futures, for Clayoquot Sound and for the people, animals, trees and plants who inhabit the area. Thus, although I do eventually come to the blockades in this chapter, one of my central points is that to take the blockades as the main site of politics would be to miss a lot of what was happening in Clayoquot. The history of Clayoquot, and its forest industry, is bound up with a colonial and a nationalist project, so much so that the challenge to this history that the FOCS and others mounted was readily recognized as such by the forest industry and the state – as the apparently hyperbolic, but in many respects rather perceptive, accusations of treason and traitors directed at the FOCS demonstrated (Rowell 1996). Intriguingly (and as I note throughout with respect to various feminist critiques of eco/feminists), the extent of the challenge that environmentalists pose is sometimes more readily recognized by the state and by industry than by critical friends and academics who risk reducing the scope of their challenge to a tame desire for wilderness. These conflicting versions of what mattered in Clayoquot demonstrate how much is at stake in the process of theorizing and the need for responsibility in the construction of stories/theories and their limits.

Empire of Wood

The conventional history of Clayoquot Sound, and of British Columbia more generally, in common with many colonial histories, is one of settlement, with one or two versions dominating.[2] Even local histories can draw on this universalist trope. One account, *Clayoquot Soundings: A History of Clayoquot Sound, 1800-1980s,* was written by Walter Guppy (1997), who had lived in the Clayoquot area since the early 1920s, when his family settled there. Guppy's history began with the "first settlers," whom he identified as Catholic priests, such as Father Augustin Brabant. In this, he discounted the arrivals who figure so prominently in other colonial histories, which tend to mark beginnings with European discovery rather than settlement. In Guppy's account, then, Juan Perez, who navigated the coast in 1774, and Captain Cook, who came in 1778 during his third voyage in the Pacific, were dismissed as fly-by-night explorers with no real commitment to the place – this the measure of a man in Clayoquot Sound, no easy place to stay and live. Many histories of the west coast begin a little farther north of Clayoquot, in Nootka Sound, with Spain and Britain fighting for the dubious honour of first contact, sovereignty, rights to navigation, trade, and access to sea otters, who were killed and stripped of their pelts, which were sent around the world to clothe and warm others, and to provide and prove status.

Although Frederick Thornberg was in charge of the Clayoquot trading post on Stubb's Island when Father Brabant and the other missionaries arrived, he does not appear to count in Guppy's history either – but perhaps being Danish, or married to a Native woman, disqualified him from the honour of first settler. *Clayoquot Soundings* hardly mentions the people, now called the Nuu-chah-nulth, who may also be settlers but who have their own, very different histories of the region.[3] By Guppy's (ibid., 7) definition, they ought to count: "If settlers are defined as being people who come to establish a permanent presence in the area, there were none as such in the Clayoquot Sound area – or the whole of the outer coast of Vancouver Island for that matter – when Catholic priests first came to establish a mission at Hesquiat in 1875."[4]

Fur trading required expeditions but did not demand permanent residence in Clayoquot. Fur traders visited sporadically until the end of the nineteenth century, but eventually immigrants arrived from China, England, Japan, Norway, and Spain to hunt seals, to fish and can sockeye salmon, and to mine for gold, silver, and copper. People kept coming. Some came just for fun; Tofino has attracted tourists for a long time. Throughout much of the 1910s and 1920s, most people arrived by Canadian Pacific steamship, on the round trips of the *Princess Maquinna* (and later the *Princess Norah*), which provided the sole transportation link with the outside world (apart from fishing boats or long hikes to Port Alberni). Eventually, people came to log when the value of timber rose in a global market. During the First World War, a heavy loss of planes resulted from the introduction of aircraft-mounted machine guns, creating a demand for timber. Sitka spruce was favoured for the construction of aircraft frames, being light, strong, and flexible, and the best supplies were on Haida Gwaii (formerly named the Queen Charlotte Islands) and the west coast of Vancouver Island. This fact provided H.R. MacMillan, future founder of MacMillan Bloedel, with the opportunity he wanted to contribute to the war effort (Drushka 1995, 105).

During the Second World War, thousands of servicemen came to Tofino and Ucluelet, the two main villages on the west coast of Vancouver Island, to prevent invasion by Japan – though people had already been arriving from Japan for some years. At the turn of the century, many had come to fish and had settled on the coast. Yet following the 1941 bombing of Pearl Harbor, the Issei and Nisei, first- and second-generation Japanese Canadians, were forcibly interned in work camps in the BC Interior for the duration of the war. Seventy-five percent of them were Canadian citizens. Most lost their homes, livelihoods, and belongings, and never returned to the coast (Kogawa 1981, 1992; see also Marlatt and Minden 2001). H.R. MacMillan was among the many people who called for the removal of all those of Japanese origin from the coast – his other contribution to the war effort.

Though Tofino was designated the terminus of the Trans-Canada Highway in 1926, the town remained virtually inaccessible until 1954, when a road was extended from Port Alberni. In 1964, the logging companies that owned the road were persuaded to open it to the public. With this development, more and more people came. First, more loggers: the need for airplane spruce subsidized the start of many BC lumber companies and marked the beginning of the "timber frontier" of the 1950s, '60s, and

'70s. Eventually, loggers were joined by Vietnam draft protesters (Haig-Brown 1996; Hagan 2001), founders of utopian communities (Brown 1995), hippies, environmentalists, and more and more tourists. Vancouver Island has attracted many utopian experiments; for as long as colonizers have been coming to exploit the natural resources and the people, others have come in search of Eden or lotus land or have created their own communities, seeking alternative ways of living.[5] Clayoquot is a challenging place, where utopia and genocide have long been entangled.

Conventional settler histories of Clayoquot Sound mask the displacements – the deaths, genocide, and ecocide – that were central to colonization. In fact, the area's history is a series of settlements *and* displacements, comings *and* goings, the goings – mainly of sea otters and fish and trees, and Japanese Canadians and those who are now called the First Nations – mostly going unnoticed. As Adele Perry observed in *On the Edge of Empire* (2001, 7), "Inasmuch as this term [settlement] suggests that nobody was there, it subtly depoliticizes the process whereby white people came to dominate First Nations territory."

Current conflicts in British Columbia, most obviously over logging and Native land claims, could be understood as evidence of the failure of previous projects – of the exploitation of "natural resources" and the "civilizing" of Native people. The strident narrative offered in MacMillan Bloedel's "We came, we sawed, we conquered" (Vidal 1993, 24) was challenged in Perry's (2001) version of British Columbia as on the fringes of empire, where the making of the province was a fragile project and where "success," which was by no means inevitable, always met considerable resistance. As Gordon Brent Ingram (1994, 43) articulated it, "In effect, the 1993 conflict over Clayoquot turns out to have been a move from an older form of blatant neo-colonialism to newer forms of over-exploitation of the land and First Nations communities that are only a little less obvious."

Traditional histories of British Columbia, and in particular of Vancouver Island, have been challenged by many, from labour historians who studied the poor conditions of workers to novelists and artists. Such challenges come most obviously from the First Nations and the land claims that are going through the Canadian court system: most of British Columbia was never ceded to the federal government through treaties, so the land claims cover the entirety of Crown land, which comprises the total commercial forest base of the province (Boucher 1997, 170). Only recently has attention been paid to Aboriginal accounts of first contact with

Europeans.[6] Some of the most visible challenges to conventional histories of Clayoquot Sound have come from the land itself. During a severe rainstorm of 10-11 January 1996, 127 landslides occurred in Clayoquot Sound. In torrential rain, mud and rock slid down mountainsides that had been denuded by clear-cutting and scarred by logging roads, a process that made them vulnerable to such erosion.[7] This kind of devastation potently undermines forestry's claims to husband and conserve the land.

Forest Husbandry and Industry Fathers

> Why shall we not let Nature continue to produce again, what she can do unaided? ... There can only be one answer, that Nature alone, after all, is not able to do what man requires; that for some reason man must aid. And the reason, the direction in which Nature fails, is that she has no idea of economy in time or space, being most wasteful of either; while man, with the constant increase of population, is forced to use both time and space more and more economically, and for the strict purposes of his needs. Forestry is the child of necessity.
>
> – BERNHARD FERNOW, 1902, QUOTED IN KEN DRUSHKA, *H.R.*

> The first great fact about conservation is that it stands for development. There has been a fundamental misconception that conservation means nothing but the husbanding of resources for future generations. There could be no more serious mistake. The first principle of conservation is development, the use of natural resources now existing on this continent for the benefit of the people who live here right now.
>
> – GIFFORD PINCHOT, 1910, QUOTED IN CHRIS CUOMO, *FEMINISM AND ECOLOGICAL COMMUNITIES*

> There are an astonishing number of people in this country who misinterpret the term forestry, and oppose any extension of forest administration because of that misinterpretation. Such people believe that forestry is the conceit of sentimental persons who desire to protect woodlands, to prohibit the cutting of trees ... Still another misconception of forestry exists in the minds of

> many persons connected with the lumber industry, who believe that a forest department would import from Europe or elsewhere absurd regulations requiring the replanting of a tree to replace every one cut, or to hedge logging operations about with killing restrictions.
>
> – H.R. MacMILLAN, 1914, QUOTED IN KEN DRUSHKA, *H.R.*

> taking out the dead wood. pruning back the unproductive. it was all a matter of husbandry, "the careful management of resources." for everybody's good, of course. a matter of course. (by definition)
>
> – DAPHNE MARLATT, *ANA HISTORIC* (1990)

The examples of Bernhard Fernow and Gifford Pinchot, founding "fathers" of scientific forestry, and of H.R. MacMillan, who followed in their footsteps, show us that history as progress narrative is bound up with history as patrilineage. Anna Tsing (1997, 257) observed that "most environmental historians take us from one great thinker to the next, humanizing the story by showing history as a family tree." It is an insidious anthropomorphism. Tsing argued that "it is the genealogical convention that makes these histories seem to work; of course we know, sons learn from their fathers." Writing about the turn-of-the-century conservation movement in the United States, commonly seen as the origin of American environmentalism, Tsing (ibid., 258) (re)contextualized this story:

> Most environmental histories obscure the earlier nineteenth-century presence of conservation by tracing the lineage through Thoreau, who was a nature writer, not a conservationist; thus it appears that the late nineteenth-century parks movement, and Muir, invented conservation. Yet in the nineteenth century, there was plenty going on about conservation; however most of it was "elsewhere" – in the European colonial periphery. Conservation was the appropriate management of other people's natural resources. What the US parks movement did was turn conservation to native uses; nature protection became appropriate for one's own people, and, indeed, essential to their individual and collective self-development.

H.R. MacMillan's distinguished forestry lineage, compounded by the conventions of patrilineage, naturalized and authorized his interventions in

BC forestry and (other) politics.[8] Fernow, a German forester who advocated the scientific management of woodlands to allow for perpetual and multiple uses, brought scientific forestry to North America. In 1886, he was appointed the first chief of the Division of Forestry in the US Department of Agriculture, and he later opened the first Forestry School at Cornell University. A number of his followers were Canadians who played substantive roles in MacMillan's career (Drushka 1995, 40-41). When Fernow resigned from the Division of Forestry in 1898, he was succeeded by Pinchot, whose family founded the Yale Forestry School that MacMillan attended. Pinchot was Theodore Roosevelt's chief of the Federal Forestry Division and a founder of forestry – and conservation – in the United States (ibid., 34-35). MacMillan first came to British Columbia in 1907 to stake timber for the summer, taking a break from his classes at the Yale Forestry School, where he was learning forest policy from Pinchot. He worked for the Continental Timber Company, one of the many syndicates that took advantage of a new form of timber lease made available by the McBride government, precipitating a staking rush (ibid., 46). He returned to British Columbia in 1912, this time as the province's first chief forester, at the request of Martin Allerdale Grainger (secretary of the Royal Commission on Forestry in 1910). With advice from Pinchot and Fernow, he established the BC Forest Service. Intending to bring BC woodlands under the management practices of Fernow and Pinchot, MacMillan believed that the Forest Service could acquire the necessary authority and funds only if the logging industry stayed at the centre of the provincial economy. Drushka (ibid., 74) wrote that

> this interpretation of the role of the Forest Service was a departure from what MacMillan had learned at Yale, and it reflected an attempt to resolve the contradiction between conservationist and industrialist ethics in a developing economy – a contradiction he would live with for the rest of his life. MacMillan's position distanced him from those of his colleagues – including Aldo Leopold, his former Yale classmate – who would formulate a preservationist argument against unrestricted industrial use of forests. Seventy five years later, the BC Forest Service would come under heavy criticism for maintaining too close and supportive a relationship with the timber industry, but 1912 was a different era, in which such a role was seen as enlightened and proper.

As chief forester, MacMillan immediately distinguished himself by selling BC lumber abroad in quantities that had a lasting impact on the provincial economy. He was at the cusp of a new era in forest "management" – and forest destruction. As Patricia Marchak (1995, 3-4) explained,

> The forest industry is a relatively recent arrival to the deforestation process. Until logging techniques were mechanized and there were mass production technologies for sawing wood, the forest industry was not capable of mass destruction. Until roads were built, access to forests was restricted. Until transportation methods accommodated long-distance carrying of lumber and then of pulp, producers and remote markets were not linked on a regular basis ... In the late nineteenth and the first half of the twentieth century, northerners established a full-scale, mass-production forest industry based on temperate-zone, coniferous (softwood) forests.

According to Drushka (1995, 77), MacMillan's desire to develop export markets for BC timber was unusual in 1912. Although export markets had been the foundation of the provincial lumber business during the late nineteenth century, when it had shipped much timber throughout the British Empire, BC sawmills had become complacent through serving the domestic market that flourished with settlement at the turn of the century. As a result, British Columbia's share of the lumber shipped from the Pacific Northwest had dropped from 44 percent to 10 percent by 1900. Anticipating that the domestic market would stabilize, MacMillan focused on the possibility of expanding exports. In 1914, he embarked on a world trip in pursuit of this goal, arriving in India late in the year. India and British Columbia were already linked by the timber industry, through Sikh immigrants who worked as loggers, and through an existing export trade to Asia.

MacMillan eventually won contracts to export railway sleepers to India in 1919. He also set up the MacMillan Export Company, which would ultimately merge with Bloedel, Stewart and Welch, becoming MacMillan Bloedel in 1952. In 1921, he signed an agreement with a US exporter to share control of the Pacific Northwest sleeper trade to India – "a legal cartel" (ibid., 123). By the 1880s, even the Himalayan forests, which were difficult to access, were nearing depletion as railroads inched their way across the country. The first Indian Forest Law, introduced in 1878, established reserves and protected forests, bringing woodlands firmly under government control. This made them more easily available for commercial exploitation while restricting access rights to peasants, who relied on them for subsistence (Marchak 1995, 186-89). In the nineteenth century, India was the centre of professional forestry in the British Empire; recast along European lines, its forestry practices had a significant influence around the world. MacMillan was impressed with what he saw. After a field trip to Assam with a forest economist, he wrote,

> The forest work the British have done in India is wonderful in the courage of its conceptions, the patience and attention it has required through long years of development, and the magnificent results, an enriched country, an increased revenue, are due only to a trained permanent staff working without political interference and the active presence of a research institute (also a forest college) which for forty years has conducted many valuable investigations. If India with its many other resources needs such a forest service, what does BC need? I only hope that we can get the government to realise that the forest revenue which this year is one half the total revenue of the province, can only be kept anywhere near its present level by expanding more of it into the maintenance and protection of the forests. Let us not be citizens of BC for today only. (quoted in Drushka 1995, 88)

This period in MacMillan's career had a lasting impact on British Columbia, its forestry practices, and its lumber trade. Europe had been importing timber from its colonies for hundreds of years, but by the turn of the century, Northern countries had begun to export to Asia and Latin America, whose forests were becoming depleted. Thus, the BC forest economy became heavily dependent on exports and has remained so ever since. MacMillan initiated industrialization in BC forestry – drawing on what he observed of how the British were introducing scientific forestry in India, including the importance of a government department of forestry, and a school of forestry. Decades later, scientific forestry and the reliance on exporting raw logs would lie at the centre of environmentalists' concerns. In the 1990s, the trade in lumber and its social and environmental ramifications were no less international than in 1900. Despite his concerns about the fate of small independent logging operations, MacMillan built his own "empire of wood" (MacKay 1982). At the same time, like so many others, this devoted family man both philandered, having the requisite long-lived affair with his secretary, and philanthropized, funding projects such as the Vancouver Aquarium – perhaps foreseeing the effects of forestry on marine life and anticipating the need to preserve species that would inevitably be endangered by intensive logging.

"Hystery. the excision of women" (and trees)[9]

> Major Nicholson! I say, George Nicholson was no more a major than your dog is! You read Nicholson's book *[Vancouver Island's West Coast, 1762-1962]* and you will notice that the Evans family are conspicuous by their absence. Well, my grandmother was the very first white woman to go to Tofino. She arrived with my grandfather in October, 1881. Back then women went with their husbands. We're a gutless bunch of nincompoops these days.
>
> – EVELYN (EVANS) SULLIVAN, 1981, QUOTED IN BOB BOSSIN, *SETTLING CLAYOQUOT*

> I learned that history is the real story the city fathers tell of the only important events in the world. a tale of their exploits hacked out against the silent backdrop of trees, of wooden masses. so many claims to fame, so many ordinary men turned into heroes. (where are the city mothers?) the city fathers busy building a town out of so many shacks labelled the Western Terminus of the Transcontinental, Gateway to the East – all these capital letters to convince themselves of its, of their, significance.
>
> – DAPHNE MARLATT, *ANA HISTORIC* (1990)

Challenges to official versions of history and attempts to rewrite them have not always come from historians. In British Columbia, novelists, poets, writers of all genres, and artists have been at the forefront of this process.[10] Joy Kogawa's first novel, *Obasan* (1981), brought the internment of Japanese Canadians during the Second World War to public awareness.[11] Challenges to particular versions of history can also come from the (fictional) wife of a history professor, as Daphne Marlatt suggested in her novel *Ana Historic* (1990).[12] Both Kogawa and Marlatt drew on existing documents and archives, but (like Walker and Woolf before them) in the absence of extensive archival evidence, and a history which told their particular stories, they developed creative ways of researching and telling stories.

Whereas Cook's exploits on his "voyage of discovery" are enshrined in official histories, *Ana Historic* creates fictional stories of women's lives in Vancouver. Their labours differ: Cook traded otter pelts and furs, but Marlatt writes about relationships between women. *Ana Historic* tells the story of Annie, who came to Vancouver with her husband, a history professor who had just found employment at the university. Annie began to research the life of Mrs. Richards, a widow and young immigrant from Britain who taught school in nineteenth-century BC. Once Mrs. Richards remarried and became Mrs. Ben Springer, she disappeared from the records, so Annie had to be creative to imagine her story, given the few official archival sources of her life (Marlatt 1990, 39). Given research on the Englishwomen who emigrated to British Columbia, the stories of bride ships, of girls from "respectable" families who were shipped to Vancouver and Victoria to become governesses and domestic servants, we have to ask whether that was to be her fate anyway, to vanish once she became Mrs. Springer. Was that story written already, before she came?[13] Marlatt (ibid., 15) wrote, "no, we don't know how she came. we know only that she was appointed teacher for the second term of the mill school's first year. a widow, they said (a safe bet), she would have been educated, she would have spoken a proper English, the Queen's they said. after all this was British Colombia, 1873."

During Annie's imaginative forays into BC history, her trips to the library to research the life of Mrs. Springer, she met and befriended Zoe. Over coffee, Zoe persisted in asking what Annie wanted from her research. She offered other trajectories for Mrs. Springer's future, and for that of Annie herself. Zoe asked about Mrs. Springer's friendship with Birdie, a single, independent woman who owned the Gastown hotel where she stayed when she arrived in Vancouver. Initially, Annie was unable to imagine this story, but it began to suggest new possibilities for her own life. While visiting Zoe's house, shared with two other women, she began to grasp the imagined possibilities of love and friendship between women and recognized them in her own unfolding relationship with Zoe.

Thus, Marlatt created a heretofore unwritten history of women in British Columbia. Like Adrienne Rich, she points to the need to imagine and fictionalize women's histories in order to envision new futures. For Marlatt,

> Official history, written history, interests me because it passes itself off as objective, documentary, as "the facts," when it usually represents the view of an elite or of one privileged historian. Even items in the so-called popular press represent particular journalists' take on what is going on. The question of what gets reported and what doesn't is a huge issue. I'm always interested in what gets left out of any official history. This is what oral history focuses on: the telling details, the daily experience of people who lived through events that later become "author-ized" as historic events – authorized in both senses of the world. (quoted in Kossew 2003, 56)

"It Happened Suddenly (Over a Long Period of Time)"

Like Marlatt, environmental activists have re-visioned the official history of British Columbia. Late in the summer of 1993, FOCS director Valerie Langer wrote "It Happened Suddenly (Over a Long Period of Time): A Clayoquot History" (1994). After a summer of protests, she chose to focus not on the camp, or the blockades, or even the story of the FOCS, but on the multiple, overlapping, uneven histories of Clayoquot. Whereas Cronon and Braun criticized environmentalists' reified notion of wilderness, the anthology in which Langer's essay appears – *Witness to Wilderness* (Breen-Needham et al. 1994) – is testimony to a more complex wilderness imaginary than Cronon or Braun appeared to credit environmentalists with. Langer organized her short essay around the annual cycle of seasons and the overlay of cycles over years, a feature common in nature writing, but not in political and national histories. In this, she followed the precedent set by writers such as Rachel Carson, who politicized the nature essay. Her history counters Macmillan Bloedel's "We came, we sawed, we conquered" narrative and offers important insights into the FOCS campaign. Yet she refused to separate out and prioritize saving the trees of Clayoquot over other issues. Rather, she demonstrated that forestry issues are embedded in other local, national, and global conflicts and times, and that challenging forestry practices necessarily involves other challenges. Avoiding any suggestion that the campaign, or BC history, advances simply toward success, she situated the FOCS among the others who took part in the conflict – First Nations, the government, the logging industry, loggers – and in disrupting the progress narrative, she resisted the idea that, as Franklin Ginn (2008, 335) put it, "national identity in many post-colonial states is predicated on nature being outside and antecedent to culture and the colonial project." She did not choose a singular point of origin for either the campaign or the destruction of the rainforest. She began "in the middle" (Tsing 2005, 2) with the reflection that ten years had passed since the dedication of Meares Island Tribal Park in 1984, an event that was an important landmark in the history of Clayoquot Sound.

Formed in 1979, the FOCS engaged in over a decade of negotiations with the BC government regarding logging in Clayoquot. Although it worked hard to gain local support for its campaign and to take part in the available lobbying procedures, it ended up in what Langer called the "'talk and log process,' where we talked and they logged the whole time" (author interview). In 1991, when the government granted a permit to log in the Bulson Watershed, the FOCS decided it had had enough. It effectively left the negotiating table and began the Bulson Creek blockades. Although the negotiations actually continued for another year, Langer remarked that the process had "failed utterly – brilliant failure after years and years of ... thousands of hours of talking" (ibid.). After that point, according to Langer (ibid.), "We just did activism the whole time. We did blockades in 1991. We did blockades in 1992. We wrote letters, and newsletters, and lobbied, and formed networks across the island, and BC." Thus, the Clayoquot blockades emerged after more than a decade of futile negotiation, as the FOCS recognized that "involving citizens in public debate can easily become a form of social control" (MacGregor and Szerszynski 2003, 19).

The year 1993 marked a turning point in the internationalization of FOCS strategy. In January, BC premier Mike Harcourt travelled to Europe in an effort to counteract the negative images of Canadian logging that were beginning to filter through. Langer, along with Garth Lenz of the FOCS, decided to follow Harcourt on his trip, to contest his account of BC logging. As Langer explained, they aimed to educate as many people as possible about the existence of temperate rainforest in Canada and "to get organizations with power and money to take on the Clayoquot campaign ... We knew we needed international help" (author interview). The trip was widely covered in the Canadian media and made a strong impression on many people, laying much of the groundwork for the blockades that summer.

In order to fund the trip to Europe, Langer and Lenz conceived a plan to create the Clayoquot Slide Show, and to show it all around Vancouver Island and in Vancouver. Small admission fees were charged and donations were solicited, with the result that Langer and Lenz managed to raise enough money. However, the slide show served other important purposes. It introduced many to the impacts of clear-cut logging in Clayoquot and mobilized much support for the FOCS. It depicted the Clayoquot conflict as a contestation over the value and meaning of wilderness, pointing to its apparently contradictory place in Canada – as central to the national identity but disappearing to support the national economy. By combining photographs of Clayoquot with a narrative that countered official MacMillan Bloedel and government stories about forestry in the area, the show presented a particular understanding of the landscape. Images of wilderness are ubiquitous in British Columbia, used by industry, government, tourism, and environmentalists. The slide show and the accompanying narrative challenged the picture postcard image of "Beautiful British Columbia" and conventional logging company wisdom. And, importantly, it included parts of the province that were not featured on any postcard – clear-cuts.

As the 1990s proceeded, forest industry publications increasingly used photographs in an attempt to insist on its version of the landscape. Controversially, Greenpeace co-founder Patrick Moore became head of the Forest Alliance, a pro-industry lobby group, and produced *Pacific Spirit: The Forest Reborn,* a volume whose photos rival any of the environmental coffee table books published then or since. In response to the challenging stories offered by environmental groups, Moore tried to wrest back some of the rhetoric for the forest industry, even describing clear-cuts as "temporary meadows," a misnomer to anyone at the peace camp, which was located in a clear-cut and was nicknamed the Black Hole as a result.

Commenting on the power of government and industry public relations, Valerie Langer reflected,

> It was very easy for them to access the entire British Columbia public with advertisements and TV commercials and a mailing to every door ... You know, that's three and a half million people in BC. They can reach them. They can change public opinion ... This is the world where media shapes public opinion and it's not hard to access BC media to shape public opinion and that's what they did and the more that we tried to work on countering public opinion in BC the less successful we were, so we just went out of BC. They don't have access to CNN – you know our government have no more access to CNN than I do – okay. And so we went to the US and we concentrated our efforts on cancelling MacMillan Bloedel customers' contracts. (author interview)

The Clayoquot conflict was often presented as one of jobs versus the environment. Jobs – "real" jobs – meant jobs in the logging industry. However, this approach ignored other, more substantive, threats to employment in forestry. Technological changes in the previous thirty years, particularly the grapple-yarder, had greatly diminished the number of workers needed to fell trees. Over a period of ten to twenty years, up to 60 percent of logging jobs were lost due to mechanization (Langer and Bate 1993, 82). Environmentalists were typically portrayed as transient blow-ins who were not locals and who had no true claim to Clayoquot. By contrast, loggers were constructed as the real local people, although many who worked in Clayoquot Sound came a hundred kilometres over the mountains from Port Alberni and beyond because their home areas had already been cut. Loggers were associated with technology, development, and progress, whereas environmentalists were seen as archaic romantics who would preserve nature and so prevent progress. Nature and progress were portrayed as mutually exclusive. Change was defined as economic progress through technological development; environmentalists were positioned against loggers, and the environment against jobs. Presenting the issue as jobs versus the environment suggested that environmentalists were unconcerned about jobs, that they were unemployed welfare scroungers, idle parasites who were unwilling to work. This image concealed the incredible *work* that environmentalists were engaged in. It was easy to portray them as urban romantics, as naive wilderness protectors. It was easy to miss the *labour* involved in trying to refigure histories of the landscape.

Logging jobs are largely men's jobs; women's unemployment in the Clayoquot area was not a public concern. In "Women Out Front in Clayoquot Sound," Valerie Langer and Jan Bate (1993, 82) wrote,

> Posters produced by industry public relations companies show a solid bunch of working fellows, chainsaws in hand, gathered round a giant, newly fallen tree. The caption above declares, "Don't Let Your Love of Wilderness Blind You to the Needs of Your Fellow Man." Indeed, the jobs so guarded by governments, unions, and industry in this debate are men's jobs. Sue Fraser, Bonny Glambeck, Maureen Fraser, Julie Draper and the host of women working to protect Clayoquot Sound who are residents of the region will not benefit from the incredible effort put into saving jobs by virtue of their sex. Logging is a bastion of male work; loggers are among the highest paid workers in Canada.

Women's employment in the Clayoquot area tended to be in the service sector, particularly in tourism. A few women had set up their own businesses, such as bed and breakfasts, cafés, a bake shop, and kayaking tours (ibid.). Much of this work was seasonal, and many survived precariously on small incomes. In a region where accommodation is at a premium during the summer, some camped out or lived in sheds while they rented their homes to tourists.

Thus, there were multiple ways in which the labour of living in Clayoquot was taken up by those women who came to Tofino and ended up making a home there. What is key, is that this "home-work," the work required to inhabit, to dwell, in Clayoquot should not be inevitably reduced to an act of (neo)colonialism or domestication. Here the scholarship of the late Val Plumwood, who did much to extend eco/feminist politics, is particularly relevant. Plumwood (2008, 141) argued that "the dissociation of affective place (the place of and in mind, attachment and identification, political effectiveness, family history, ancestral place) from the economic place that is such a feature of the global market is yet another manifestation of the mind/body dualism that has shaped western tradition." Key to Plumwood's (ibid., 141) work, which calls for an ethics of place, is her assertion that it "must help make visible north/south place relationships, where the north/south pole operates as a correlate of (various kinds of) privilege." This ethics of place, in contrast to some environmentalists' reification of a sense of place and commitment to preservation of special places, reveals that the ability to protect one's own place is yet another marker of privilege. This distinction is useful for tracing the work of the FOCS; arguably, the FOCS has been engaged in a project that insists on the ethics and politics of place.

Canada: Brazil of the North?

> Canada is the Brazil of the North. Brazil is losing one acre of forest every nine seconds. We're losing one acre every twelve seconds.
>
> – "COLLEEN McCRORY," THE GOLDMAN PRIZE (1993)

> The Latin visitors were surprised when Jim Darling, a Tofino biologist and one of the directors of the Biosphere Project, talked at length about the lack of field research into the rainforests of BC. The journalists said after years of seeing Canadian, American and European biologists traipsing through the tropical rainforests of South America, they just assumed that they already knew about the forests in their own countries.
>
> – RICHARD WATTS, "B.C.'S HANDLING OF LOGGING ISSUE HAS LATIN JOURNALISTS SNICKERING" (1992)

> Canada has been extremely active in advancing international co-operation to devise global solutions to these problems, though it has done little as yet to match its activity on the international front by cleaning up its own front yard.
>
> – FEN OSLER HAMPSON, "CANADA'S INTERNATIONAL ENVIRONMENTAL AGENDA" (1990)

The FOCS campaign went beyond a critique of clear-cut logging to expose Canada's role in global deforestation and the international timber trade. In pursuit of this, it adopted the Brazil of the North campaign, which had been initiated by the Valhalla Wilderness Society. In juxtaposing logging in Brazil with logging in British Columbia, the FOCS contested easy narratives of Canada as an advanced nation and undermined popular perceptions of deforestation – that it occurred mostly in developing countries and that the main site for concern was the Amazon basin.

Canada's stance on forestry became clear at the United Nations Conference on Environment and Development (the Earth Summit), in Rio in 1992, the year before the Clayoquot camp. It was the first country to ratify the Convention on Biodiversity, but at the same time it was one of the Northern nations that blocked a legally binding forest treaty, even though forests, especially temperate rainforests, are a crucial component of biodiversity. Third World countries at the Earth Summit opposed the treaty because it failed to acknowledge the ongoing deforestation in Northern countries, including Canada, and held the South primarily responsible. It was at the Earth Summit that forests and forestry emerged as a specifically international issue on the political agenda of governments. In the end, the "Non-Legally Binding Authoritative Statement of Principles for a Global Consensus on the Management, Conservation and Sustainable Development of All Types of Forests" was signed, but because it was not legally binding, it could not force action against damaging forestry practices. As Toner and Conway (1996, 122) observed in their review of the internationalization of Canadian environmental policy, "In some cases, Canada's prime motive for being involved with an international institution will be defensive, to protect its economic interests with respect to environmental and economic decisions made internationally or in other countries. Canada's interest in an international sustainable-forestry agreement is, at least in part, to avoid being boycotted."

Canada later successfully appealed to the United Nations to designate Clayoquot as a site of global environmental significance under the UN Biosphere Programme, with the result that it became a UNESCO Biosphere Reserve on 21 January 2000. This development was popularly understood to contribute to its protection, but, though the biosphere reserve designation did provide a commitment to sustainable development, it was not legally binding, and it permitted industrial logging (by Interfor) to continue unabated.[14] When the biosphere reserve was inaugurated at a ceremony in Pacific Rim National Park, it was protested by FOCS members, who dressed as endangered species, as well as by members of the Tla-o-qui-aht First Nation.[15]

Ecocitizens or Ecoterrorists? The Nature of Environmental Citizenship

> Democracy gives us the choice of which machine to vote with. Or choose which brand of razor blade we'd rather cut our throat with.
>
> – SUNG ON THE CLAYOQUOT BLOCKADE ONE MORNING IN LATE 1993

In the same years as the Clayoquot peace camps and blockades, Environment Canada, the federal ministry of the environment, published *A Primer on Environmental Citizenship* in an overt effort to teach Canadians the meanings of acceptable environmental conduct and to cultivate environmentally responsible citizens. As the *Primer* (Environment Canada 1993) explained, "Environmental citizenship is about contributing, about doing your part. It means being informed and getting involved. It means believing you can make a difference. It means caring for Canada." The *Primer* (ibid.) elaborated that "the goal of the Environmental Citizenship initiative is to help provide Canadians with the means to make environmentally responsible decisions. The information you have accessed is from a primer that has been written as part of this initiative, in order to foster the development of an environmentally literate citizenry."

The Clayoquot blockades were probably not anticipated by Environment Canada. Yet the blockades and the subsequent mass trials offer a particularly potent site for examining the challenges involved in bringing together the terms "environmental" and "citizenship," not least because the origins of the term "environmental citizenship" have been credited to the *Primer* itself (MacGregor and Szerszynski 2003, 8). Yet civil disobedience as a particular way of practising and enacting environmental citizenship was hardly what Environment Canada had in mind when stressing the importance of taking action: "Express your support for spaces and species, whether in words or actions." That people were arrested and jailed for what could be seen as practising their environmental citizenship suggests that only certain environmental behaviours (such as recycling) are acceptable and to be normalized, acting as a reminder that juridical power can function as the correlative of biopower when citizens refuse to conduct themselves appropriately (Rutherford 1999, 41). Although, "for greens, civil disobedience is a potentially legitimate expression of active citizenship" (Doherty and de Geus 1996, 5), it remains controversial, as emerged in the Clayoquot campaign. Clayoquot activists were labelled "an environmental terrorist movement," "tree-saving zealots engaged in environmental guerrilla tactics," "extremists," and "terrorists" (Rowell 1996). Even academic work on environmental citizenship often focuses on rights and responsibilities – on rights to a clean environment and responsibilities for private behaviours for the common good – and on a renewed participation in democratic conversations about protecting the environment (Barry 2000, 126). In this, it commonly neglects civil disobedience and "illegal" actions taken to secure a more sustainable world. So it is perhaps unsurprising that the *Primer* had nothing to say about turning to non-violent civil disobedience when voting, or writing to government representatives, fails to resolve environmental problems. In contrast, the Clayoquot camp provided an opportunity for transformative learning through daily workshops on non-violence and civil disobedience, history, philosophy, and practice, covering exemplars such as Gandhi and the suffragettes. So too did the experience of the blockades and the ensuing arrests and trials (Moore 2003).

Although Foucault rarely addressed the environment explicitly, his work on governmentality and biopolitics has proved fruitful for those who seek to understand the disciplinary and normalizing aspects of environmental policy and politics. This has led to the coining of such terms as "green governmentality" and "environmentality," which could be understood as critiquing neo-liberal notions of environmental citizenship, such as that manifest in the Environment Canada *Primer* (see Darier 1999; Demeritt 2001; Braun 2002; and Agrawal 2005). Although Foucault's biopolitics, his concern for life, was largely anthropocentric in its attention to the control of aspects of *human* life, a number of theorists have extended it to include *all* life forms (Darier 1999, 23). Thus, the transition from sovereignty to governmentality, which Foucault identified as coterminous with the beginning of an era of biopower, as the entry of "life" into history, can, in the context of environmentalism, also be thought of as the entry of nature into history. In this way, a Foucauldian approach enables the resituating of efforts to cultivate environmentally responsible citizens in the context of the emergence and history of the state (Darier 1996, 593). This bringing together of Foucault and environmentalism has proved productive in illuminating the regulatory aspects of state involvement in the cultivation of green civic virtues. For instance, Darier (ibid.) argued of Canada's 1990 Green Plan (a $3 billion environmental action plan to guide federal policy for five years), which predated the *Primer* by three years, that it did not express the sovereign will or desire of the people, but rather aimed to create an environmental citizenship, a particular environmental subjectivity, and to discipline Canadians into becoming an environmental population. Environmentality, then, points to the federal government's uneven application of the demands of environmental citizenship and its own uneven implication in the governing of conservation, while at the same time it continued selling the natural resources of the country. As David Demeritt (2001, 445) wrote, "The notion of governmentality points to their [nature and the state] mutual constitution. The construction of nature as a territorially delimited quantity of resources went hand in hand with that of the state as the agency charged with conserving them." Whereas environmentality helps to explicate state involvement in environmental regulation, it has also been used to highlight the role of environmental non-governmental organizations in controlling environmental policies, especially in the Third World (Luke 1999). However, and following Arun Agrawal (2005), I am more interested in the shifts in subjectivity, particularly around citizenship, among those who were arrested for acts of environmental citizenship.

Many who engaged in non-violent civil disobedience and who blockaded the logging roads at Clayoquot did so because they had become frustrated by the limits of voting, lobbying, and letter writing, a disenchantment that was fuelled by their ever growing anxiety about the logging, which persisted throughout. As Gordon Brent Ingram (1994, 17-18) put it, representing the issue as a conflict between corporate logging and conservation of one of the largest remaining areas of coastal temperate rainforest "prevents us from looking at the ways in which decisions were made for the Sound's use, decisions which arose from personal ambitions, ineptness, dysfunctional bureaucracies, and corruption." BC Green Party leader Stuart Parker, who was arrested during the blockades, pointed to the ability of the forest industry to subvert the democratic process in the province: "With their massive ad campaigns, their well-placed political donations, their commissioned polls, their tactics of intimidation and fear, the transnational forest companies have controlled this province's governments, economies and people for generations" (quoted in MacIsaac and Champagne 1994, 177).

The 1993 peace camp and blockades were precipitated by the Clayoquot Sound Land Use Decision (CSLUD), which was announced on 13 April

1993 by Premier Harcourt. The CSLUD permitted clear-cut logging on 74 percent of the Clayoquot Sound land base. It shattered the remaining hopes of Harcourt's supporters, who included both environmentalists and logging industry employees (not mutually exclusive categories in any case), that the province's first "socialist" government in many years would rectify the "sympathetic administration" of the Social Credit regime, which had pandered to the huge logging companies and ignored the concerns of environmentalists and loggers and their unions.

Speaking on the telephone with me, Clayoquot arrestee Barbara Taylor recalled her reaction to the announcement of the CSLUD. It was vividly imprinted in her memory: While listening to the radio during a long car ride home to North Vancouver from Iowa, she heard that the NDP would permit the clear-cutting of Clayoquot Sound. Fired by a sense of betrayal, she sent a three-page letter to Harcourt. I heard variants of this story many times from others. The late Irene Abbey, then in her eighties, was arrested, like Taylor, at the blockade. During my interview with her, she discussed the announcement of the CSLUD:

> Mike Harcourt had a press, a news release, but instead of having it in Victoria, as you would expect, he sneaked off to some little spot near Tofino. As a result on TV, I saw Ian Gill, who was a great journalist, and he was at a phone booth by the side of the road, and he was saying to somebody, "Can anybody tell me where the news release is being held, because it starts in ten minutes and I don't know where it is." One of their best journalists can't find where the news release is! A sneaky thing to do, and I thought, by golly, I'm going to do more than I have ever done – we have been betrayed. I wrote to Harcourt and told him so, but, no, he never answered me. He was just a disaster, we had finally got an NDP [New Democratic Party] government and they had betrayed us worse than the other guys, and then, I knew that way back in 19 – whatever it was, forty or fifty years ago, Robert Sommers, had accepted bribes from MacMillan Bloedel and gave them cutting rights, and they still have them. And he went to jail for four years because he took the bribe money, and they still have the cutting rights – illegal! And yet I had become a criminal, because I stood on the road for ten minutes, the logging trucks got a full day's work in that day, and I was in contempt of court – this is really a disgrace.

Like many others, Irene Abbey challenged the notion of what constituted criminality and of who was to be considered a criminal. Learning of the alliances between MacMillan Bloedel, the government, and the Royal Canadian Mounted Police (RCMP), campaigners became more determined to assert their own positions. Many connections were discovered: arrestees were taken to Ucluelet jail in a MacMillan Bloedel Forest Tours Bus, which the company kindly loaned to the RCMP; the government owned substantial shares in MacMillan Bloedel; and it was revealed in court that company employees and the RCMP had exchanged the names, addresses, and photos of protesters. MacMillan Bloedel's long history of violating the terms of its licences was made public, as was the insignificance of the resulting fines that it paid. In an effort to contest who was a criminal and what counted as criminal activity, the Clayoquot Rainforest Coalition produced an information sheet titled "MacMillan Bloedel's Criminal Record," which listed sixty-six violations of provincial and federal environmental laws. In this context, and as the trees continued to fall, many felt they had no alternative but to protest. Moreover, what should have been treated as minor infringements and taken as civil cases by the logging company against the protesters ultimately became criminal cases. As arrestee Maurice Gibbons recognized, "What began as a confrontation with a logging company removing old-growth rainforest ended up as a confrontation with the Supreme Court of British Columbia over our violation of its injunction" (quoted in MacIsaac and Champagne 1994, 92). Thus, the Clayoquot blockades emerged at the end of over a decade of "participation" as the FOCS recognized that "involving citizens in public debate can easily become a form of social control" (MacGregor and Szerszynski 2003, 19), and as it endeavoured to find ways to refuse to participate on the terms established by the provincial government.

For Valerie Langer, the legal response to the campaign entailed "the disenfranchisement of those who had, up until the Clayoquot Blockades, considered themselves full members of Canadian society" (quoted in MacIsaac and Champagne 1994, 189). In countering this and in explaining their civil disobedience, protesters invoked discourses of citizenship. They denied claims that they were bad citizens by insisting that their actions sprang from a sense of responsible and active citizenship. For many, the decision to get arrested was an achievement born out of a struggle with their own concepts of what good citizenship might mean in practice. Most arrestees had no previous convictions. Kim Back discussed her struggle over being arrested: "I'd never been arrested. I'd never even had a ticket! You know, I never jumped the stile in a subway or anything like that, you know! I am a fine upstanding citizen" (author interview). Arrested with more than eight hundred people in one of Canada's largest acts of non-violent civil disobedience, this "fine upstanding citizen" was sentenced to electronic monitoring. Arrestee Carol Johnson argued in court that civil disobedience was a manifestation of good citizenship:

> Civil disobedience is evidently not well-understood in these parts. The rule of law is not flouted but affirmed by the symbolic gesture of concerned citizens peacefully expressed. Democratic society is not threatened but revitalized by such considered expressions. Those of us who stood on a public bridge accessing public lands did so as a profound expression of the seriousness with which we take the obligations of citizenship. (quoted in MacIsaac and Champagne 1994, 127)

Yet the meanings and practices of citizenship are not universal. The campaign, the blockades, and the subsequent court proceedings were sites of contestation over national identity and belonging, over the meanings of a specifically *Canadian* citizenship. Because wilderness has played such a key role in the creation of Canada, activists who opposed logging practices and the logging industry were also challenging the foundations of the nation, potently illustrated when their adversaries called them "enemies of the nation" and "traitors." For Catriona Sandilands (2000, 181), "wilderness takes up a concrete position as national origins and essence, from which any departure is not only unnatural but un-Canadian," a comment that would resonate among Clayoquot activists. For many protesters, a central paradox of the campaign was that wilderness, so vital to Canadian identity, was what brought them into conflict with key Canadian institutions: the RCMP, the courts, the provincial and federal governments. They posed their challenges in a number of ways: through acting as witnesses to the destruction of Canadian wilderness; through the process of arrest, trial, and sentencing for their acts of non-violent civil disobedience in blockading a logging road; through learning about Canada's role in the global trade in timber and related forest products; and through informing themselves about Native politics and land claims. This was true for many activists and for members of the public who learned of the campaign through the media. Indeed, the sense of public outrage at the treatment of arrestees by the courts and prison system, as well as at the destructive logging practices revealed by the activists, was a key aspect of the campaign's success. As Valerie Langer observed, "Comfortable Canada suddenly faced oppression when its citizens challenged the status quo. Friends were being arrested, silenced, jailed. The sort of politicization that resulted is the kind no group can organize. No pamphlets or educational video could have done for the grassroots environmental movement what the Supreme Court of Canada and the BC government did" (quoted in MacIsaac and Champagne 1994, 189).

Activists consistently engaged creatively with national symbols, such as the national anthem "O Canada" and the Canadian flag. Although some objected to singing the anthem, others rewrote it, singing "Our home *on* Native land" in place of "Our home and native land." As Eva Mackey has observed "nationalists' representations of nature also reflect a central conflict over whose 'native land' the settler nation of Canada now occupies" (2000, 125). Some protesters sang the official version but inflected the lyrics with new intensity. A group who came from Denman Island, off the east coast of Vancouver Island, brought a decidedly wordy placard: "I stand not in contempt of court but with respect for my country and its ancient forests," as it attempted to resist the labels of lawbreakers and bad citizens. In singing the anthem while being arrested, group members insisted on the weight of the words, calling the Canadian government to account for not taking its own sentiments seriously enough, for not "doing *its* bit":

O Canada!
Our home and native land!
True patriot love in all thy sons command.

With glowing hearts we see thee rise,
The True North strong and free!

From far and wide,
O Canada, we stand on guard for thee.

God keep our land glorious and free!
O Canada, we stand on guard for thee.

O Canada, we stand on guard for thee.

There was also dissent over the use of the Canadian flag (as well as other national flags). Yet in invoking national symbols, activists often reappropriated them to new ends, all the time stretching their received meanings. Wearing T-shirts and carrying placards that featured the Canadian flag, with its lone maple leaf, they demanded that "Canada" stand not just for the few leaves that would remain after the forests had been clear-cut. Texts inscribed on T-shirts insisted that "My Canada includes Clayoquot

Sound." Other T-shirts extended "My Canada includes" to trees, old growth, bears, and the marbled murrelet, in recognition that multiple "Canadas" might be at stake in the campaign. Such gestures can be understood, following Sandilands (1999, 89), as arguments for a distinct subject position for "nature," which can be seen as "a radical move in the direction of democracy." Elsewhere, Sandilands (2000, 185) noted, "One is left with a rather more anarchic terrain in which one may well find abjection in the company of a human other and a friend in the company of a tree, thus opening the struggle for Canada to a wider variety of interactive/intersubjective possibilities than is currently possible." Sandilands's (ibid., 181) observation that "wilderness takes up a concrete position as national origins and essence, from which any departure is not only unnatural but un-Canadian" would be familiar to many at Clayoquot. The symbolic role of wilderness in Canada meant that in challenging logging practices and the logging industry, activists were, in fact, also challenging the foundations of the nation.

In engaging in non-violent civil disobedience and in being arrested, activists in Clayoquot were also drawn directly into conflict with the state. In this way, activists may be seen in a long line of agitators in British Columbia. As Adele Perry (2001) observed, women in the province have not always acted as a "civilizing influence" in quite the ways intended. Although coming to Clayoquot Sound felt like coming home for so many women, the ethical work of being at home there required and demanded significant labour, homework of a rather different kind from that anticipated by Environment Canada's *Primer.* To collapse the campaign into a defensive effort to save wilderness, or into the work of media-savvy protesters is to miss much of the politics and strategy that were involved in the work that the Friends did in drawing the international and the global into the Clayoquot campaign. This political and ethical work is more usefully understood as a manifestation of Plumwood's "ethics of place," which pays attention to north-south relationships.

5

"It Was Like a War Zone"

The Clayoquot Peace Camp and the Gendered Politics of (Non)Violence

God, it [the Clayoquot Peace Camp] was ugly. I couldn't believe how hideous, the horrid shot-rock gravel roads that the forest company had put in, in order to make it a "tour site" or something at some point. Their plan was that they were going to do these little forestry tours or something through this business, so they had crushed granite, which was in big horrible sharp chunks, I'm sure you remember it, and I was just appalled, just appalled, at a sense of poverty. I was struck by how many people that were there, came across to me as homeless, or economically marginal, or well below the line, that, I was really struck by the marginality of it, and grim, grim was definitely what comes to mind, and then as one walked through the camp and got to sort of, the other side, where the little tent pods sort of meandered up, it took on a whole different kind of character, and it felt more like a little village, and there was more of a, a, kind of – it felt friendlier, but I was struck by, just the grimness and oh, dogs everywhere, and just thinking, oh, this is really bad, I have to spend time here, and I was, I was impressed that people had spent the whole summer in those conditions, that you know, wow, this is, people really getting down with the basics, that it was a huge level of commitment to spend time in this, and that that rapidly wore off and it became normal, within a couple of days of being there. On my first trip out I stayed a few days, and I was also struck by how it became a home-like environment, and I got used to the, the kind of ambience of the place, and felt that, from the beginning, I felt that it was a completely appropriate setting for it, that it be in a completely devastated place, made no bones about what was going on there, and sitting in the camp, and looking up at the, at the mid-slope roads that you can see from the actual camp, and how they are washed out, it was like, well, this is what we are here about, and but, really, really struck me as ugly, and very west coast, very much, such a different terrain there from [Denman Island, where] we live in this sort of moderated gentle inner channel area, and out there everything is so much more, almost vertical, and raw, so all of those things combine to make it quite a jarring sort of setting.

– WENDY POPE, AUTHOR INTERVIEW (1996)

And once I got to the camp, I was completely enthralled, it was just the most amazing experience, and the most amazing experience for my son, to see all these different people from all walks of life, all sharing food, all sharing chores, working as a tribe, and it was just amazing. I think that was the greatest experience of my life to see, you know, garbage getting picked and dealt with, food preparation, feeding thousands of people and shared food, it was just amazing. It really renewed my faith in humanity.

– JUDITH THOMPSON, AUTHOR INTERVIEW (1996)

But it just reminded me of that kind of like slap-it-together, kind of, it's not like, it reminded me of summer homes I'd had, you know, like with me and my kids, we'd had to camp several occasions in the summertime and it – it struck me as incredibly well organized, especially since it was stuck out in the middle of this hell-hole, you know, it felt, it felt really smoothly, smoothly put together and – the whole food thing! By itself, I mean can you imagine having to feed all those people, you know? ... I mean, and it was great! I mean, the food was delicious, and it was hot, and filling, and really tasty and nutritious, and it happened! You know? And I loved the respect that everybody had for each

> other, and the feeling of, there was an instant community ... Anyone, who stepped onto the property, right, was like part of this family ... And that was really nice. Lots of smiles, and you know, easy to talk to people that you'd never seen before and find out their stories.
>
> – KIM BACK, AUTHOR INTERVIEW (1996)

> In the early days, the camp was very, very exciting, extremely exciting, and by the end it was very sad, it was very, very sad.
>
> – MIRIAM LEIGH, AUTHOR INTERVIEW (1996)

There are many stories of the Clayoquot Peace Camp. These are just some small extracts from interviews with women who had been there. Women I spoke with, such as Wendy Pope, Judith Thompson, Kim Back, and Miriam Leigh, gave vivid, detailed, and varied descriptions of the camp. One of the most powerful stories about the camp was about eco/feminism. In the first chapter, I mentioned some of the popular and public manifestations of claims about eco/feminism at the camp: the award-winning documentary film *Fury for the Sound: The Women at Clayoquot;* Jean McLaren's account of daily life at the camp, *Spirits Rising: The Clayoquot Sound Peace Camp, 1993* (McLaren 1994); Betty Krawczyk's *Clayoquot: The Sound of my Heart* (Krawczyk 1996); Chris Lowther's *A Cabin in Clayoquot* (Lowther 1997); and the *Vancouver Sun* article "Eco-Feminists Run 'Peace Camp' at Clayoquot Sound" (S. Bell 1993). The peace camp brought the politics of eco/feminism in the campaign sharply into focus. It was the site where an emergent interest in eco/feminism in (some of) the members of the Friends of Clayoquot Sound (FOCS) was most visibly and tangibly articulated and put into play. The camp was said to be based on eco/feminist principles, expressed through a commitment to the philosophy and practice of non-violence and on a consensus process; a Code of Non-Violent Action was displayed at the camp. Yet the eco/feminist basis of this code, as understood by many of the organizers, was not necessarily transparent to participants. This lack of fixity about eco/feminism, what it was, and what bearing it had on the camp permeated the summer's events. The peace camp was the place in the campaign where eco/feminist politics emerged, were fought over and contested, and perhaps disappeared. This chapter examines the manifestation of eco/feminism at the camp and also uses this as a site through which to revisit fraught disputes in feminism about feminist peace activism as a particular manifestation of essentialism, perhaps even *the* exemplary essentialism.

Peace Camps, Essentialism, and Eco/Feminism

The Clayoquot Peace Camp offers a productive site for revisiting the often passionate and troubled debates in feminism about pacifism, non-violence, maternalism, and essentialism – and for suggesting alternative ways of understanding these conflicts. Feminist peace activism of the 1980s, exemplified by protests at Seneca Women's Encampment in the United States and Greenham Common in the United Kingdom, has been contentious for many feminists. Such activism was, and is, often perceived as the epitome of essentialism because of the assumed circulation of maternalist discourses that relied on arguments about women's role as childbearers and -rearers, whether this related to women's biological or social role as mothers, to characterize women's pacifism. As discussed in Chapters 1 and 2, from the mid-1980s on, the labelling of any political or theoretical position as essentialist crystallized many problematic struggles among feminists. The feminist peace activist can be seen as a mythical figure of this era, the archetypal essentialist, yet feminist environmental activists now often attract the contempt once directed at feminist peace activists. Eco/feminists are derided for allegedly suggesting that women's role in caring for children may leave them more inclined to care for the planet too. Eco/feminism appears to have reignited anxieties that have lain dormant given the commonly assumed demise of feminist peace activism. Such a reading of feminist peace activism and eco/feminism collapses a complicated politics of non-violent activism into manifestations of maternalism and then further collapses maternalism and essentialism. In this process, essentialism has been invoked to disavow feminist peace activism and eco/feminist activism. Throughout this book, my aim is to trace the ways in which essentialism has been produced as one of the key concepts on which certain feminisms are seen to flounder, particularly activist eco/feminisms, and how essentialism is implicated in those narratives of feminism that characterize the early 1990s as the moment of the end or death of feminism.

For instance, in the field of feminist peace politics, the opprobrium surrounding essentialism is so great that those who are interested in feminism and non-violence pursue their interests under the more acceptable label of "feminist anti-militarism." However, this focus on the military as a locus for thinking about feminism and non-violence results in some losses. The litany of activisms, from Greenham and Seneca to Chipko and beyond, which have been so central to feminist accounts of the 1970s and 1980s, have been replaced by a much more troubled citation of women in the military, girl gangs and ladettes, female suicide bombers, and notable individuals such as serial killers Myra Hindley, Karla Homolka, and Aileen Wuornos. Any number of events, such as the invocation of mothering (and fathering) on both sides of the US debate regarding the war in Iraq, have revealed the limitations of any easy linkage of motherhood and non-violence. Motherhood can be, and has been, invoked for many ends, acting as a stark reminder of the futility of automatic recourse to a peaceful maternalist politics. That some feminists have turned to theorizing violent women, in recognition that women are not innately peaceful, should come as no surprise, though anyone who has spent time at a women's peace camp could testify to this. Although the desire to reposition feminism as a discourse of power rather than victimhood does make a certain sense, we should be alert to the limitations of any simple reversal of dualisms. It is also the case that the focus on particular and spectacularly violent women, such as Hindley, Homolka, and Wuornos, suggests a liberal individualizing and psychologizing imperative, which constitutes a shift from the intersubjective, relational, institutional analysis that typified the ethos of many peace camp activists. Although theorizing violent women and critically interrogating the problems of essentialism have been necessary, this does not obviate the need to think through the categories of, and relationships between, non-violent activism, essentialism, and narratives of feminism. In exploring gendered conflicts at the camp, I suggest that women's violence in contexts where non-violence is prized suggests that feminists could theorize violence much closer to home, in ways that could prove more productive than turning their gaze outward.

Rather than reading the camp as essentialist and hence retrograde, or anti-essentialist and thus sophisticated and clever, or strategically essentialist and therefore intentionally risky, I explore how the categories of essentialism and woman have been invoked and to what ends.

The Friends of Clayoquot Sound and the Clayoquot Peace Camp

The Friends of Clayoquot Sound (FOCS) formed in the late 1970s to protest clear-cut logging and turned to direct action in the late 1980s. It blockaded on a number of occasions, at Sulphur Pass in 1988, Bulson River Road in 1991, and Clayoquot Arm Bridge in 1992. Then in 1993, it blockaded the Kennedy River Bridge and set up the peace camp to enable supporters from outside of Clayoquot to come and join in the civil disobedience. The immediate impetus for the camp and blockades had been the announcement by BC premier Mike Harcourt of the Clayoquot Sound Land Use Decision on 13 April 1993, which would allow logging on 74 percent of the land base of Clayoquot Sound.

Initially, the FOCS had little organizational structure, and volunteers chose their own level of involvement, focusing on whatever issues interested them, resulting in activism that waxed and waned. During the mid- to late 1980s, the FOCS grew as a number of people moved to Tofino and became involved in campaigning against logging. This infusion of energy renewed levels of engagement. By the early 1990s, organizational changes had taken place, of which the most significant was a commitment to feminist principles, defined as a turn toward consensus process and the philosophy and practice of non-violence. However, this change from an ad hoc approach to organization was contentious. Valerie Langer, one of the key campaigners during the 1993 camp and blockades, described the meeting where these issues came to a head:

> *Valerie:* We had a meeting – like we're always having meetings – and two of the men at the meeting just dominated the whole time. And after, finally after an hour – both of whom, you know, are men that I admire, and I am still friends with – they just dominated the meeting and finally [another FOCS director] said, "You know I'm just tired of listening to you two have your private conversation. I've had something to say for the last half hour and don't get a chance to get a word in edgewise, and I think it's time regardless of whether you have more [to say]" – 'cos they kept on saying we have another thing to add – "regardless of whether you have something more to say, I think you should give somebody else a chance to talk." And this created a furor, and you know there was this, "Whenever I have something to say I shouldn't be shut up, that's creativity." What came out of it was that the meeting kinda blew up, and these two – one guy said, "Oh, I didn't realize," and the other just said, "Ah, I can't stand this 'feminist stuff,' I'm getting out of here," and left.
>
> *Niamh:* Had anybody mentioned feminism at this stage?

> *Valerie:* He did. Because the people who were saying they thought it was time they stopped talking and somebody else get a chance too, were all the women in the group. So, we decided at that meeting that we should begin consensus-model decision making and have a process for running our meetings. You know, anarchy was fine until you didn't get to speak, until somebody else always got a chance to speak, and you couldn't get a word in edgewise. So we started at that time organizing ourselves as a consensus decision-making organization with feminist principles, and it happened to be the feminists in the group who were willing to stick it out, and, I shouldn't say, yeah, the feminists in the group, some of whom were men. Yeah, who said, "Yeah, I agree totally and this is – let's start looking at how we do it." So the organization shifted again there towards having a structure and a process.

Although feminism was not initially mentioned when the challenge was made to the "right" to dominate discussion space, this challenge was identified, and rejected, by one of the men as "feminist stuff," whereas the other accepted it as legitimate "feminist stuff." In Valerie Langer's account, then, feminism was not tied specifically to women; on consideration, she included men in her understanding of the "feminists in the group." This account of feminism in the FOCS can be situated historically in the context of a politics in which feminism is no longer understood as being solely about women or gender. In the wake of criticisms, specifically from women of colour, white feminists had to rethink the subject of feminism and the process of doing feminist politics. Thus, Langer's version of feminism in the FOCS was not structured around any particular subject or identity as women. Instead, it presented feminism as a way of understanding power relations as a matrix of interlocking systems of dominance and subordination, and as a politics that demand attention to process and practice. There are two related issues to note here. First, this articulation of feminist principles did not rely on a recourse to "woman" as a foundational and necessary category for feminist politics; feminism was open to men too.[1] Second, the refusal of feminism as an identity politics, the destabilizing of the subject of feminist politics, opened up and made way for a version of feminism that focused on the process of political activism.

In Langer's account, this new emphasis on organizational structures fed into ideas about the practice of activism. Whereas the 1988 blockade at Sulphur Pass had involved little more than gathering a number of people and a boat to reach the site, the 1991 Bulson blockade was more carefully planned. Putting into practice the principles of non-violence and consensus decision making, FOCS activists held off the loggers for six days and were arrested on the seventh. The emphasis on consensus decision making and the higher level of organization and safety may have contributed to the successes of the Bulson blockade, but accounts of the peace camp reveal that consensus, non-violence, and the relationship between these and feminism remained contentious and were never fully resolved.

In preparing for the summer of 1993, the FOCS drew on its experiences of the 1991 and 1992 blockades. A crucial aspect of the 1993 blockade, which was to be much bigger than that of the previous year, was the need to attract large numbers of people from other parts of Vancouver Island, Vancouver, and the rest of the province. Tofino is a small village with a population of under two thousand (that is, in winter; the summertime population of this popular tourist resort has been estimated at up to twenty thousand). Many Tofino locals had been arrested in the previous blockades and were still serving probation, which forbade them to be near active logging roads. The camp was necessary to accommodate the expected visitors, providing them a place to sleep before attending the early morning blockades and an opportunity to participate in workshops on non-violence and civil disobedience.

During my interview with her, Langer described two key preparations the FOCS made in planning the camp, drawing on its previous experiences:

> We spent enormous amounts of time building a network of support across the country and in our bioregion, Oregon and Washington, and we held several non-violence civil disobedience trainings in Vancouver, Nanaimo, Victoria, and here. We got a number of non-violence trainers to agree to come up and be part of the camp and to train trainers. So we did a trainers' training before the peace camp began. You know, one of the things we learned from the 1991 and '92 blockades is how vulnerable you are when you are asking people to come in and help you. You get everybody and you don't necessarily get the people who you would prefer to be with. You get the people who have some "inner desire" to be part of a movement – whatever the hell that movement is – and so we knew that we were going to have to define what the movement was, because people just wanted to be part of, but they weren't necessarily committed to non-violence, and we wanted to make sure that the movement that we created, were creating,

> was dedicated to the philosophy of non-violence, and philosophy and practice of non-violence. So we, we spent a lot of time getting trainers trained and peacekeepers trained; and then the other key thing was to make sure that there was good food. Two key parts to a good blockade, non-violence training, good training and organization, and food. And so we spent a lot of time making sure that we'd set up the camp so that it was focused around how you would participate in the non-violence of the actions, and how you would participate in preparing food communally, because that brings people together and creates a sense of community.

The FOCS stressed the importance of non-violence by prominently displaying the Code of Non-Violent Action on the notice board, beside the area where meals were eaten and meetings held, and on the way to the kitchen, where people queued for meals. The code read,

1 Our attitude is one of openness, friendliness and respect toward all beings we encounter.
2 We will not use violence, either verbal or physical, toward any being.
3 We will not damage any property, and will discourage others from doing so.
4 We will strive for an atmosphere of calm and dignity.
5 We will carry no weapons.
6 We will not bring or use alcohol or drugs.

Although the non-violent philosophy and practice of consensus decision making were widely and visibly enacted at the camp, their links with feminism were not necessarily clear. It is true that many FOCS members and their supporters were committed to feminist politics and that some were interested in exploring eco/feminism. And, as Langer suggested, a number of male activists were feminist or pro-feminist, and many women mentioned the importance of male allies. Furthermore, networks of women played significant roles, with women calling on their female friends to help organize the camp. Langer also referred to the importance of connections with women in other BC environmental associations. Nonetheless, in the context of the camp, the meaning of eco/feminism was not always clear, and it was sometimes vigorously contested. That the ideals of non-violence and democratic decision making might be linked with eco/feminist principles was not necessarily obvious to those who understood feminism as about women's issues, narrowly defined.

Perhaps the most visible reference to eco/feminist politics was in the welcome handout, quoted below, which people received on arriving at the camp. However, the fact that someone devoted an evening to crossing out the word "ecofeminism" on a bundle of welcome handouts suggests just how contentious eco/feminism was, or became as the summer progressed. I return to this later. Under the heading "Intent of Clayoquot Peace Camp," the handout stated,

> This is an action base-camp. We are here to bear witness to the destruction, to peacefully resist that destruction, and to educate ourselves and the public about these issues.
>
> You are welcome to participate in the day-to-day running of the Camp, and planning of actions – everyone is a participant. We ask that you volunteer each day to help the Camp run smoothly.
>
> We use a consensus process based on feminist principles. We believe that sexism, racism, and homophobia are forms of oppression which are linked to the oppression of Nature. We strive to make Camp a safe space, free of oppression. (quoted in McLaren 1994, 76)[2]

Here, then, is an indication of what feminism meant to the camp organizers. Notably, feminist principles were not restricted to issues of gender or sexuality, but were defined in terms of linking sexism, racism, and homophobia with the oppression of nature. Connections with other "isms" and oppressions can be inferred here; the handout's short list might not be exhaustive. Thus, feminism was not just about women, but about challenging all oppressions, including that of nature, and about understanding the links between them. This feminism, with its emphasis on the oppression of nature, might be understood as eco/feminist.

Feminism at the camp was defined through processes – consensus and non-violence – and an understanding of the interlocking nature of many oppressions, rather than through the construction of an identity politics. This form of feminism was not unique to the camp; indeed, it was an expression of the directions and concerns of Western feminism in the early 1990s. Basing feminist politics on the category of woman was no longer tenable – and not only for deconstructive feminist theorists. The question of how to sustain activism across differences is crucial for feminists. In a small community such as Tofino, where the politics of race, class, gender, and nature, to name only a few, are everywhere visible, understanding feminism as only about women, and then as only about certain women, remains impossible. The peace camp expression of eco/feminism was not a simple reification of "women" and "nature," but rather a critical perspective on hegemonic constructions of women and nature. It involved a commitment to non-violence and to the creation of an alternative community. Campaigning, and bringing about social change, was not solely about blockading logging roads, but also about envisioning, and practising, different ways of living. At the camp, feminism was invoked as a particular understanding of power and politics, embodying an alternative system of values and a decision-making process that challenged the hierarchical power relations embedded in conventional decision making such as that which produced the Clayoquot Sound Land Use Decision. Notably, the maternalist discourses purported to be dominant in women's environmental activism and at women's peace camps were significantly absent. This version of eco/feminism disrupts any simple recourse to essentialism as a way of reading the events that occurred in Clayoquot Sound during the summer of 1993.

Yet though I offer this recuperative account of eco/feminist activism in the FOCS and at the camp, I do not intend that it can be read as a definitive and resounding rebuttal of any and all charges of essentialism in eco/feminism generally, or even at the peace camp specifically, or to suggest that therefore eco/feminism is never essentialist. However, it should also be clear that I am not interested in evaluating various instances of eco/feminism, here or elsewhere, as essentialist, and so dismissing them. Nor am I interested in providing a reading or interpretation of any events as anti-essentialist and therefore superior. Rather I want to examine the moments where the category of "woman" (and "man") might appear and be invoked to examine to what ends and with what effects. I have pointed to the intentional politics of (some of) those who organized and participated in the peace camp. But it would be a mistake to assume that this intentionality was (or ever could be) sustained at all points through the summer by any one individual, and a huge assumption to suggest that this ethos on the part of the organizers might have completely permeated the camp and all those attending. Thus, in the next section I move on to describe how the consensus process and non-violent philosophy of the camp did come under stress. Interestingly, many of these tensions manifested themselves as conflicts over the place of feminism, and gender issues, at the camp. In this sense, feminism was reduced by some participants to being about "gender" only, and the wider implications of feminism, as a complex critique of interlocking systems of domination, and a commitment to a process of consensus decision making and non-violence, as understood by (some of) the organizers and participants, were lost. Then I turn to examine differing accounts of an incident that reveals the gap between eco/feminist intention and daily practice at the camp, and the conflicting versions of eco/feminism being articulated in these accounts. Just as my account is not intended as a definitive rebuttal of essentialism, neither is the following intended as an account of the failure of eco/feminism, or as evidence of the inevitable reiteration of essentialism in eco/feminist politics, but rather as a reminder of the ongoing and continual work that politics requires and demands. The "problem" of essentialism cannot be resolved once and for all, but rather demonstrates the importance of "a continual reflexivity over how the borders which sustain conceptual entities such as 'woman' and 'women' are constituted through acts of exclusion or othering" (Ahmed 1998, 91). My attention to conflict at the camp is intended in the spirit in which Dana Shugar (1995, x) offers her study of conflict in separatist women's communities "in hopes that our conflicts will become one part of, rather than the end of, our community" – that is, that rather than being implicated in a narrative of the end of feminism, that stories of conflict will constitute part of the ongoing and continuous story of feminism.

Gendered Conflict at the Camp

The camp aimed to be inclusive – the commitment to the philosophy and practice of non-violence extended throughout life at the camp and was not just restricted to the blockades. The camp was intended to be a safe place, or at least as safe as possible. Meanings of safety and security extended beyond common-sense notions of security, beyond safety from intruders, beyond being peaceful on the morning protests, to include a notion of personal safety and security, safety within the camp, safety also from the potential infractions of other participants. However, the consensus process and non-violent practice of the camp did come under pressure, particularly as tensions grew between FOCS members who had planned and organized the camp, and those who lived there for the summer. Furthermore, as time progressed, levels of exhaustion increased for everyone involved.

Many of these tensions manifested as conflict over the place of feminism and the attention given to gender issues at the camp. In this sense, some participants did perhaps reduce feminism to being about gender, with the result that its wider implications, as understood by the organizers through consensus and non-violence, were lost.

Once it opened, the camp took on a momentum that the FOCS had not foreseen, as people came for a night and stayed for weeks or months. Initial hopes – that perhaps 1,000 people would come and that perhaps 200 would be arrested – were wildly surpassed. In the end, what happened that summer exceeded everyone's dreams and some people's nightmares. On 1 July, opening day, over 250 people came to the camp, and 13 stood on the road to be arrested. At the height of the summer, the blockades averaged 300 or 400 people a day, and by summer's end, over 12,000 had passed through the camp. In addition, the FOCS organizers, who lived in Tofino and worked at the office, had not anticipated the sense of community that developed in the camp. Jan Bate addressed this topic during my 1996 interview with her:

> But also being a [FOCS] director, you know, we were having a lot of meetings, long days that went from, if you were on the blockades, from 4:30 a.m. to midnight or whatever, til the meetings were over, trying to make decisions. And there was a lot of flak around, like the office was in town, and it was pretty much a half-hour drive out to the Black Hole where the camp was, and the camp became a community which we hadn't designed, you know. We had designed, like a cattle show, or something, where people would file in, go through their non-violence training, go out on the road, and either get arrested or not get arrested, and go away. But that didn't happen, a lot of them stayed, and this was community, they learned about consensus, they learned about non-violence, they learned about communal living, and they got in touch with clear-cuts. They were living in a clear-cut and occasionally taking trips into the old-growth forest as well, when the clear-cut got to be too much. So with those people, with all of those people becoming community there, they more and more had distaste for those of us that were running our butts off in the office come walking in there and say, "Yeah, here is information from the outside world, and guess what, here's a new rule." They didn't appreciate that at all, and trying to work with consensus, which was actually modified consensus, like we, as the Friends of Clayoquot, were legally responsible for the safety of those people.

As Bate pointed out, once they moved into the camp, people rapidly became involved in the nitty-gritty work of building community. This included volunteering in the kitchen to prepare food for hundreds of people three times a day, and staffing the gate, both to welcome new arrivals and to act as peacekeepers in case of threats and violence to the camp. Discussion "circles" often took place after breakfast and supper, when people gathered in a circle or a number of concentric circles. The circles had no chairperson or leader, but rather one person would act as facilitator for discussion and for reaching consensus about the direction of the campaign. His or her role was to help the group work through the agenda and reach consensus. Jean McLaren (1994, 69), one of the key non-violence trainers at the camp, explained consensus:

> Consensus is a way of making group decisions by mutual consent of the participants rather than by a simple majority vote. The goal is to reach a democratic decision that everyone can live with ... This does not mean that everyone is always in perfect agreement, but that the decision doesn't violate anyone's principles and that those who don't agree don't disagree enough to stop the group from taking an action. Good use of the consensus process produces intelligent and creative decisions by incorporating the best thinking of everyone involved. Everyone has an interest in implementing a group decision because all have participated in its formation. Consensus used properly relies on synthesis and cooperation rather than competition for reaching group unity.

The morning circles typically reflected on the blockade of that day. Usually, the morning's arrestees had returned from Ucluelet, where they had been "processed" at the jail – fingerprinted, photographed, asked to sign an undertaking not to return to the logging road, and then released to await trial. They would share their thoughts and experiences with the circle. The evening circle focused on developing agreement about the intention for the next morning's action. Workshops on non-violence took place after lunch. Thus, meals and circles, where consensus processes were used in planning, formed the basis of daily life at the camp. However, as Jan Bate pointed out, in the end, everything was subject to negotiation. That problems arose in connection with the consensus model is unsurprising, as Bate suggested, given that it was actually a modified form of consensus. As she pointed out, the FOCS made many decisions about the camp before it opened, and certain aspects of it, such as non-violence, were not negotiable. In *Truth or Dare,* Starhawk (1987, 183) reflected that problems arise with consensus because groups do not clearly understand what it is. She argued that "consensus is not so much a decision-making process as a thinking process of the group mind. It is a way to consider problems, not just a way to choose between alternative courses of action." Starhawk (ibid., 187) also stated that consensus is sometimes not an appropriate decision-making process: "When deep divisions exist within a group, or when members don't value the group's bonding over their individual desires, consensus becomes an exercise in frustration." Although workshops on consensus and non-violence were offered every day throughout the summer, attendance was voluntary and not everyone availed themselves of the opportunity. In any case, a few hours of workshop training might not have been enough to undo a lifetime's exposure to "democracy," majority rule, and the investing of authority and responsibility in a small number of elected officials. At the camp, no topic was safe from dispute – even the duration of the workshops was not off-limits. As Jean McLaren (1994, 8) noted,

> I had designed a format that I thought would work and it took about three and one half hours. Some people in the FOCS office thought that we should do five hour workshops, especially on the weekends, but I found that this just would not work. People were not prepared to sit on the hard ground, through a workshop that long, when they were so emotionally charged about being arrested and tired from going to the blockade.[3]

One participant even referred to the shorter workshops as the McDonald's version, an insult indeed in environmental circles and an indication of just how controversial the issue was.

Objections to consensus and the failures of consensus were often gendered, as Fireweed, who also led the non-violence training workshops, pointed out during her interview with me:

> There were men who were extremely uncomfortable with the whole process of determining as a circle, that the whole non-violence civil disobedience, that we could plan our actions for the following day and not just reinvent the wheel. And ad nauseam we would have to explain to people that this isn't about tying yourself up to a tree, this isn't about lying down in front of the trucks as they're rolling towards you, it is not about independent actions. This is about a team effort, this is about people working together – and it really got tedious.

Describing an evening circle in which gender issues came up, Jean McLaren (1994, 33) wrote that "some people thought that we should not even be talking about this." As McLaren (ibid.) was facilitating the circle, she decided that the subject needed to be addressed, saying, "I have been in many movements for over forty years, and never was there any time to discuss gender issues. We always seemed to have to put them aside for another time. I am willing to sit here all f*ing night if you want to talk about it." McLaren (ibid., 34) recounted that she herself was particularly upset when, after she spoke to the circle about her own abuse, a male participant said that he was tired of hearing women talk about being abused. The circle went on long into the night, with energy flagging for many. No consensus or resolution was ever reached, as was the case for many ongoing issues that summer.

Gender issues also arose in connection with food at the camp. The Clear-Cut Café won universal praise for producing excellent vegan food for hundreds of people daily and for contributing to the creation of community by enabling everyone to eat together (see Gaard 2002 for an excellent overview of vegetarian eco/feminism). It provided a space for some volunteers to meet and get to know each other, but much of the kitchen work remained gendered, as women tended to prepare the food, and most of the cooks were men. Furthermore, the first cook, who was employed by the FOCS as part of the strategy to provide good food throughout the summer, was said to have sexually assaulted a young woman at the camp. When she disclosed this, she was offered support from other women who held a women's circle where they shared stories. Women also asked their male allies to hold a men's circle, where questions of masculinity and power were to be discussed, although there was some fear that the men's circles were sometimes a place to moan about women, rather than to constructively challenge gender roles.[4] Ultimately, the cook was asked to leave, and the kitchen was staffed entirely by volunteers for the rest of the summer. It is unlikely that this sexual assault was an isolated occurrence: according to one interviewee, other young women who had been at the camp told her about being sexually assaulted and harassed by men there.[5] Attempts to deal with sexism at the camp were never entirely successful. As Miriam Leigh remarked during my interview with her,

> The issue of sexism was just never, ever put to rest. There were these men that were just anarchists, and they never really got the thing about non-

> violence, and they wanted it their way, or no way, and, I'll never forget one night in the circle, it was the night Joanna Macy [author of *Despair and Empowerment in the Nuclear Age*] was coming, every night they always had a talking circle, and the issue of the treatment of women in the camp came up, and this very disgruntled man talked about how he was raped over and over by the corporate bosses, and this woman stood up, she didn't just stand up, she just streaked straight up out of her seat, just like a rocket, right into the middle of the circle and just looked at this man and said, "Don't you ever use the word rape as to what's happened to you, you know, you don't know what rape is" – which could be considered, y'know, men can be raped, but, it was like emotionally the men never got it, what it was like to have been a woman in this society. They were like on the Robert Bly trip, the drums at night never stopped y'know. It would have been great if these guys had just got up and painted themselves blue or something, and just gone and gotten it out of their spirits.[6]

This man was identified by many as obstructing the feminist ethos of the camp, and he was nicknamed Iron Ron, after Robert Bly's *Iron John* (1992), a key text of the mythopoetic men's movement. Iron Ron objected to the creation of a women-only space at the camp and tried to veto the women and children's blockade. Given that in the consensus process, decisions can be vetoed by any individual, this was a serious challenge. In her examination of consensus, Starhawk suggests that vetoing should be rare and that objections to action should be discussed in the group in an attempt to reach resolution. Although Iron Ron recognized some aspects of patriarchy, such as the domination of young men by their older or wealthier counterparts, he did not develop an analysis of how patriarchy is gendered, how women experience it, and how women and men might be located differently within it. In realizing that the political was also personal, his analysis was individualistic, not perceiving that the personal was also political. Most importantly, he failed to understand his own role in patriarchy, that his behaviour in camp – demanding attention, obstructing consensus on the need for a women's space and the women and children's blockade – represented for many the perpetuation of dominance and hierarchical ways of decision making. Some women later suggested that Iron Ron should have been asked to leave the camp as he was unwilling to accept its feminist basis. However, this did not transpire, because the commitment to consensus prompted people to engage with him and with others who supported him. Many long hours were spent on this fruitless exercise, which caused endless frustration, drained energy, and prompted conflict regarding whether it should be pursued at all. Ultimately, disagreement arose among women and feminists over whether dealing with gender issues was appropriate in the context of the camp, with some people arguing that its purpose was to save trees – not to resolve gender conflict.

An apparently more trivial issue – drumming around the campfire at night – also caused problems. This was a recurring theme in the interviews. Most of the late-night drummers were young men, who were drumming to "the beat of the mother's heart," which led many to retort that "this particular mother needed her sleep." Drumming shortened the already brief time for sleep and was thus a crucial matter. A regular day began with the 3:45 a.m. wake-up call and often did not end until midnight or maybe later – though many drummers did not get up to go on the blockades. Lack of sleep was central in fraying nerves and tempers, and it became an increasing problem as the summer wore on. Despite continuous efforts to engage with the drummers, the drumming continued all summer with little relief. During her interview with me, Wendy Pope mentioned that many people came to the camp just because it was a place to hang out for the summer and that they also provided a focus for the media:

> In a way I feel like the credibility of the whole thing was harmed somewhat by that, it gave the media that aspect to really focus on, which they did mercilessly, and yet I can't find it in myself to criticize those people, to say it's not, that anything that they were doing was not okay. It was terribly annoying at times. I mean people that wanted to drum all day and people were trying to have conversations about stuff that had importance to what was going on politically, and so there was an ongoing sense for me of a kind of young and restless energy, and that without people like Jean McLaren, who was camp mother basically for the whole summer, it wouldn't have been anything like as successful as it was.

The necessity for continuous intervention over drumming was one instance where a maternalist dynamic displayed itself, with various "camp mothers" having to intercede and "nag" recalcitrant young people, most of whom were male. Furthermore, drumming not only damaged the credibility of the camp by giving eager reporters something to focus on, but more importantly, it harmed the camp's internal dynamics by requiring huge amounts of time and energy to deal with people who, for whatever reasons, resisted its dominant ethos. Conflict occupied more space than anticipated and sapped many people's energy in the process.

Alistair Bonnett's analysis of masculinity and the mythopoetic men's movement is useful here. Bonnett (1996, 275) identified group drumming as a common activity of the movement, along with wilderness retreats, group incantation, initiation rites, the (re)creation of male rituals, and the invocation of age-based male hierarchies. He discussed research that saw the men's movement as a response to socioeconomic changes that had affected middle-aged, middle-class men: losing status and security in the workplace (in part because of its increased gender, racial, and sexual heterogeneity), they had perceived this development as a threat to masculinity. However, Bonnett argued against any simplistic reading of the fact of economic change and pointed to the ability of many middle-class individuals to develop a political consciousness that is based on a "post-materialist" interpretation of socioeconomic change. Bonnett (ibid., 278-79) suggested that "the absence of a working class or non-white mythopoetic men's movement says more about the nature of contemporary political activism and cultures than the absence of a workplace transformation or crisis of masculinity among these groups." He identified the assertiveness of the mythopoets as reflecting their fear of male powerlessness and their contradictory experiences of power – they experience both power and a sense of its loss in their relationship to feminism. When the peace camp drummers claimed that they were drumming to the beat of the mother's heart, they were drawing on what Bonnett (ibid., 288) described as "primitivist discourses of identity and landscape to naturalize its adherents' contradictory experiences of power," discourses that displace "gender from social or economic consideration onto the mythic terrain of nature." For Bonnett (ibid.), "This process may legitimately be understood as representing a 'backlash' against feminism." He (ibid., 279) added that "this experience of challenge, with its concomitant loss of masculine authority, should not be exaggerated. Feminism has unnerved many men but its power must be placed in the context of competing masculinist agendas and initiatives ... For men, being threatened by feminism in a masculinist society, is to encounter conflicting sensations of power."

It is useful to reflect on differences between young men and older men at the camp. Bonnett's analysis may be helpful in understanding an older man such as Iron Ron, who felt that he'd been repeatedly raped by the corporate bosses and who perhaps had encountered conflicting sensations of power in the workplace. Nonetheless, it does not fully account for the situation of the young men. Their presence at the camp may have signalled a rejection of the corporate world, in keeping with Bonnett's recognition of the importance of post-materialist values. However, their relationship with power and corporate bosses might have differed from that of older men, as they were not yet caught up in this world. It is worth noting that Iron Ron and the young drummers occupied different spaces at the camp – Iron Ron dominated the evening circles, with his arguments and objections to feminism and his insistence on his own oppression in the world. According to interviewees, the drummers rarely attended the circles, taking their place at the centre of the camp only after discussions had ended and most people had headed to their tents to sleep. Unlike Iron Ron, they did not argue and demand attention in the circles, though as Wendy Pope noted, their "young and restless energy" pervaded the camp. They provoked annoyance rather than outright anger and fury, as Iron Ron did.

The camp's feminist ethos facilitated transferring skills from older to younger women, and it encouraged women to come forward and use their skills. Many young women took leadership roles, and many women were empowered through the process of passing on skills to other women. The contributions of older women such Jean McLaren were valued, but so too were those of younger women. Arguably, a similar culture did not exist for the young men.[7] However, such ready acceptance by women of all ages of the responsibility for maintaining the camp can easily be understood within codes of femininity, where the work of maintaining daily life for others is encouraged, and where such practices are not part of the common expressions of masculinity. During her interview, Wendy Pope reflected,

> I felt, when we had the evening circle, and these incredible eighteen-year-old, nineteen-year-old women, stepped forward to facilitate and did a phenomenally good job, I was just wowed, we need lots more of these, and then I would look again at the young men, and I didn't see that kind of energy coming from that quarter. I saw it from the young female quarter much more, much more strongly, the leadership was really coming from the women, that's quite clear, and I hadn't really seen that, until I just looked at that now – that it was the women, I mean I hadn't seen that before, I hadn't seen that it was so much not the guys. That wasn't so apparent to me.

So, older men such as Iron Ron evidently did not conceive of themselves as role models or mentors for younger men; nor would the young men have found him an appealing mentor. Nonetheless, it is also important to note that many men made dedicated contributions to life at the camp, often working quietly in the background. Valerie Langer commented on this fact and recalled that they were never a focus of the media; people were so surprised by women's leadership that they failed to notice the men who also worked hard throughout the summer. Similarly, the claim was often made that two-thirds of the arrestees were women. Yet when I checked the names of arrestees in the FOCS office and undertook the admittedly crude exercise of assigning gender on the basis of the names, I found that the numbers of male and female arrestees were about equal.

Whereas some young people, especially young men, were the focus of censure because of their lack of constructive contributions, others prompted distress because, frustrated with the tactics and pace of the camp, they planned alternative actions of their own. During her interview, Miriam Leigh described her initial frustration with this group and her gradual reconsideration of its tactics:

> It seemed to me there was a split happening in the camp between just going out there every day and standing in the line and blocking logging trucks that go in anyway, we're not doing anything, we need to escalate and do more Earth First! stuff, so that segment of the camp never got it that the whole purpose of the protest was to be inclusive, that to do Earth First! type stuff, you need to be young and able bodied, which totally precluded all the older people or the not so able-bodied people ... So yes by the end of Clayoquot I was so disgusted, yeah I have to say that I was so disgusted with environmental politics and what I saw as the treatment of women in the environmental movement, and also starting to question myself how effective was peaceful Quaker-style non-violent action. Maybe these young people were right, maybe it was time to take to sugaring the gas tanks, and building platforms, and mucking about that way. I had started to question my own beliefs around non-violent direct action, and because of circumstances in my own life, I had been assaulted by this man, who was my partner in giving non-violence workshops, at the end of our relationship.

These issues are important not just because they caused conflict, but because of their impact on the long-term life of the camp. The constant tension drained energy, with the result that, near the end of the camp and in the immediate aftermath of the summer, many people were exhausted and experiencing burnout. Exhaustion easily leads to bad decisions or to enacting decisions badly.

Other problems arose because FOCS members had difficulty keeping up with the exponential growth of the campaign, as Valerie Langer mentioned during her interview:

> 1993 was the first time that we actually hired people as staff and the FOCS changed and here we are with computers, a fax. You know we didn't even, when I started in 1988 we had a phone and a photocopier; it was dial phone too, no push button anything, so we had an answering machine. We had an answering machine, that was it, and we got computers in 19 – when was it? I think we got, in 1992, we got a computer which nobody knew how to use except for Dan Lewis, and in 1993, we bought another computer. We had two computers and we started getting lessons on how to use them, and then we bought a fax machine, so it really wasn't pretty close up to winter 1992, close on '93 that the FOCS even had a fax machine.

The difficulties intensified once the campaign was divided between the FOCS office at Tofino; the camp, which was eighteen kilometres away; and the blockade at Kennedy River Bridge, a few kilometres from the camp. The FOCS was unprepared for the communication and transport issues that arose once the campaign was spread across the three sites.

As the summer progressed and the camp grew beyond expectations, security became both increasingly pressing and hard to manage. Spread over a large area, the camp had many points of entry, even if some were not very accessible from the road, and monitoring all of them was impossible. New people were constantly coming and going, and infiltrating the camp would have been relatively easy – indeed, there were suggestions that this had occurred on a number of occasions.

The Gendered Politics of Peacekeeping and Non-Violence

> Plus there were bears! ... Yeah! There are no predators on Denman, and to be in this little nylon [tent], it was kind of freaky for me too. So I'm like, oh my god, like I've got chocolate! We have to eat the chocolate, right now! We were like, we eat the chocolate, we do not enjoy the chocolate, it was really an exercise! Right? [munching noises]. Eat the chocolate! And then E– was like, "You know, I have sleeping pills." She says something about, she's

on sleeping pills, I was asking her all kinds of questions about sleeping pills, 'cos I don't medicate myself, like, you know, it's not my thing, so I ask her all these questions and she's feeling really uncomfortable, like I'm going to pass a judgment on her, and finally I say, "Can I have one?" So I take a sleeping pill, 'cos I'm like, I know I am *not* going to sleep. So we get into the tent and all night long I think the little boy next to me is *the bear!* Every time he snores – I'm laying awake, you know, tense like a board, and my support person Louise is in the same tent, and she's laying awake, tense like a board, and I keep thinking, "Oh, I wish Louise was awake," and she was thinking, "Oh, I wish Kim was awake." And then at one point, could hear some kind of commotion, someone came to our tent and said, "There's an intruder in the camp – be alert." And I went, "Okay, I'll be alert" [snore].

– KIM BACK, AUTHOR INTERVIEW (1996)

I tend to feel safe in the world, so, part of that is what we carry with us. I am not victimized, I don't get victimized, I've never been victimized, and lots of women have, and it's hard to feel – well, I'm sure I've been victimized, am I not a woman? – but not in the way I think of lots of people have, so it's hard for lots of people to feel safe anywhere, but I found the camp a really safe environment. My only fear, my only fear, my only anxiety of a safety issue, was the harassment from the Ucluelet folks that were driving by and hurling bottles, and hurling abuse, and part of what made me feel safe was, was the peacekeeping, that they got met with every time, but I was always worried that someone would be at the gate who would react to it, or that they would find some hook to get in, or that they would just be drunk and rowdy and come and hurt someone.

– WENDY POPE, AUTHOR INTERVIEW (1996)

Well, things that became out of hand, there was one woman who, when the camp was raided on by young yahoos from Ucluelet – I think there was, in that raid nobody actually got hurt, but she was severely frightened and later diagnosed as having post-traumatic stress syndrome, and she, for, in the week before that

> had been trying to get through to the coordinators to say that this is not safe here, like "We are not safe here and we need to deal with this," and after that had happened to her, she was still trying to say that, she was saying, like, "I can't be here any more, but you guys need to look at this," and we didn't have the structure and we didn't have the time and space built into the structure to really look at how do, how can you make it safe.
>
> – JAN BATE, AUTHOR INTERVIEW (1996)

The question of safety at the camp was a serious one – whether the danger came from the bears that roamed near the kitchen or the loggers who feared the loss of their livelihoods. As Jan Bate suggested above, safety was a major concern for the FOCS. Throughout the summer, threats came from logging supporters who drove by the camp, often late at night after the Ucluelet RCMP had gone off duty at 3:00 a.m. Sometimes they were drunk. Whereas threats from loggers and their supporters were to be anticipated, those that arose *within* the camp itself were less expected, and there was no structure in place to deal with them.

This is particularly true of a violent incident that occurred late in the summer, when a number of local men entered the camp late one night and attacked a woman in her tent. In examining two accounts of this event and its aftermath, I keep in mind a comment from Gatens (1996, 30), who notes that a genealogical critique focuses our attention not on "the chronological process of what happened in time, but the historical record, the narrative account of what happened in time."

The first account was written by Jean McLaren, whom I interviewed. The second is from a long in-depth interview with Fireweed, who was the focus of the attack. McLaren's published version echoes that given in her interview, and it provides a sense of the status of the incident and shows that stories about it circulated widely within and beyond the camp. Furthermore, the two narratives can be read as if in conversation with each other and with the various versions of the event that circulated publicly. I quote extensively from Fireweed, not least because she spoke at length about the subject, which became the lens through which her experience of the camp was refracted. Many people mentioned the episode, but for them it was only part of their story of the summer; for Fireweed, it was central, and she and her partner left the camp because of it. In addition, it resonates because both Fireweed and McLaren were key figures at the camp; both were identified with the articulation of eco/feminist politics, and both taught workshops on non-violence and peacekeeping.

McLaren's (1994, 15-16) version of the incident, published a year after the camp, went as follows:

> The next night, a Saturday, some of the same drunks came back and harassed the people on the front gate at which time the police were called on the radio hook-up that Kevin had rigged up in his trailer. A few of the rowdies crawled through the rough terrain at the far eastern edge of the Black Hole and threw some stones at the tents of three women who were in an isolated part of the camp away from the others. There were also some children in this area as it was quieter than other places, and felt to be quite safe. When the stones were thrown, one of the women yelled for help which came in less than three minutes. The peacekeepers did a great job and the police arrived and removed the intruders. One of them was the son of the local president of SHARE BC the anti-environmentalist group located in Ucluelet and funded by [MacMillan Bloedel]. We had heard that the police had trouble with him before but it also seems that no charges were laid. This intrusion upset many in the camp and caused two key people to leave and never return. However, I felt that it brought us together in a way that had not happened before. The men seemed to sense how vulnerable some of the women felt and we came to a better understanding. It was interesting and frightening though, to find that many outside attacks, either verbal or physical, on the camp from anti-environmentalists were against women.

Fireweed's account, taken from her interview with me, was lengthy and detailed:

> I was scared shitless ... It was all I could do to sit up and zip up the side of the tent, like, right, that's going to protect me, and I tell you I was afraid to breathe, and all of a sudden I start to hear the voices, and I realize, that this was, that there was no bear, this was a group of men ... You know, you just started imagining the worst, praying that it was all paranoia, that it would all go away and stop, but it didn't, they finally got to the tent, started hurling, swearing, cursing, swearing, hurling insults, and heavy objects started hitting the side of tent, and I started to scream, and I screamed and I screamed, and I screamed, until help finally came. I was so hoarse, I lost

my voice from screaming, that's how long I screamed for help … And I'm screaming, *"When are you going to listen to the women?"* and I lost my voice from that. I was just hysterical, I was hysterical, because I thought I was going to get killed … I was hysterical, and later, I was attacked for this, how I had traumatized this young woman. I'm the one who was in the fucking tent, thinking I'm going to get raped or shot, lying there in the dark, and it's like these people had no idea how to handle my fear. I guess I was supposed to be sweet and demure and just cry quietly. Well, no, I was fucking hysterical, and that's how I handled it, the trauma, you know, other people might handle it differently … And I'm naked in the tent, and meanwhile these guys are still all around us in the woods, they have not split … There is running going on, you don't run, rule number one, you don't run. [My partner] jumps up on a stump with the flashlight to try to flash it around, and I am yelling, and this is when we are finally out of the tent, and I'm yelling, "Get down, get down, they are going to shoot you," you know, because here he is flashing, and he turns off the flashlight, and these guys [the intruders and the peacekeepers] are hurling insults back and forth, and [one of the peacekeepers] cursing and swearing at these guys, and they are cursing and swearing back. It's like this is not what you call peacekeeping, guys, this is not guerrilla warfare you know, and I am standing there shaking, and because I just threw my clothes on, and [my partner] says, "Come on we gotta get out of here," and these guys [the peacekeepers and the intruders] are running around in the woods and yelling at each other, so I'm standing there alone. I have just been through this trauma and there is nobody there to, "Fireweed are you okay, it's okay you know," give me a blanket, anything you know, and so I'm finally, I finally yell for [my partner] and say, "Come on, come on, like you said, let's go now, let's go now" … So I finally, we tried to get some sleep, and I couldn't sleep, this woman was sobbing across the way, and I just had to get up and go and deal with that, and, I didn't have to, but I wanted to, I couldn't bear to hear her, and it was great, it was the best thing that could have happened, and there were these four other women, and they were sitting there and they are crying, just saying like, how fucked it was, those guys [the peacekeepers] at the front gate, and this one woman had tried to warn them that these people [the intruders] were on the way. Her boyfriend had left the camp, they'd had a fight, she'd gone driving down the road after him. Anyway, she'd saw the people coming in the cars to attack the gate, and she had witnessed them, she was the one that had complained and had wanted the cops called the night before, they ignored, and she was actually leaving

> the camp and saw these guys, she came back to the camp, at her own risk, to tell the guys at the front gate that these guys were coming and was backlashed, again everything else, apparently it was utter fucking bedlam at the front gate.

Both accounts agreed that men entered the camp and attacked some tents, but their portrayals of the camp differ significantly, and both relied on and produce very different versions of feminism. McLaren presented the incident as an intrusion that was successfully dealt with by the peacekeepers. Although two key members left the camp, her story is nonetheless celebratory. For Fireweed, the story is about the failure of peacekeeping and non-violence – it is a story of internal failure.

Fireweed's account began with the danger posed by the intruders but quickly shifted to the threat as internal, to conflicts over the meanings of peacekeeping and non-violence, and the feminist basis of both. For her, peacekeeping was reduced to security and to a paternalistic sense of responsibility for protecting women from physical violence, rather than a desire to maintain peace and well-being. In a slide from peacekeeping to policing, to malestream models of "conflict resolution," the peacekeepers argued over how quickly they had come to her "rescue," and the people who diminished the trauma of the occurrence suggested that she had overreacted and was hysterical, reverting to stereotypes of women's emotions.

Other women who were attacked at the camp did receive support. Fireweed's response to violence did not correspond with the cultural conventions of female victimhood, as she said, "I guess I was supposed to be sweet and demure and just cry quietly" – and perhaps some other women's responses did. People were accustomed to a Fireweed who was capable, strong, and independent; they had no idea of how to handle her fear. Furthermore, she insisted that both the attack and the general response to it were not private, individual issues. Because they were related to camp organization, they needed to be made public and should be understood as connected with the efficacy and gendered nature of its peacekeeping. Although feminists have theorized that language such as "hysteria" has been used to police women's emotions, to disempower them, and to dismiss their analyses of and experience of oppressive power relations, Fireweed's friends, who were also feminists, reproduced these ideas. Her hysteria was reduced to personal emotions and was not perceived as a political resource. That it might have been an analysis and critique of the camp was not seen.

Clearly, understandings of peacekeeping differed, as did the implications of its limitations. Perhaps it was easier to deal with other women who had been attacked and whose responses could be aligned with cultural expectations of women's victimhood. Fireweed's response, enabled precisely because of a feminist framework, revealed the gap between intention and capacity at the peace camp and was not at all welcome, even though it could be seen as echoing the intentions of FOCS feminism, with its emphasis on non-violence and consensus. In McLaren's account, the camp was attacked by "a few rowdies" who threw stones at tents. The peacekeepers did a "great job," the police removed the interlopers, and the upsetting experience was contained. The moment of loss, in which two important members depart, was followed by a coming together of those who remained. The incident prompted a positive change in attitude, which was defined in terms of gender: some men gained a better understanding of the vulnerability felt by women, and McLaren herself realized that women are often the focus of outsider attacks. However, her account reproduced and reinforced traditional gender roles, of women as defenceless and of men as protectors. The language, of "naughty boys" rather than of "violent men," is crucially important, as is the way it engenders the subjects and objects of peacekeeping. There was no suggestion that the conflict over the meaning of peacekeeping and non-violence was addressed, or indeed, that it even existed. Dissent was erased in this narrative – unlike in the case of Iron Ron, whose criticisms were consistently addressed. Although some may have regretted the persistent engagement with Iron Ron, the commitment to working with him is instructive, suggesting further layers in the gendered nature of conflict. Ultimately, "community" was secured by the exclusion of dissenting voices and conflictual positions, rather than via the process of working these through or through continuous attention to the boundaries of community.

This event and its divergent renderings highlight a moment when the struggle for alternative praxis floundered over gendered issues – when the camp was revealed as not a safe space for (some) women, not solely because of outsider attack, but because of conflicting notions of gender and peacekeeping. Despite the good intentions of those at the camp, the incident makes clear that it is sometimes easier, intentionally or not, to revert to essentialist discourses, that in the face of difficult events and emotions, maintaining a more nuanced feminist politics and practice can become too arduous. Challenging dominant cultural norms and their effects requires ongoing work, even, and perhaps especially, at a peace camp based on feminist principles. In this fraught moment, the more complicated gender politics that some had wished to espouse were lost in the recapitulation of essentialist discourses of women as vulnerable and men as protectors. These discourses crept back in when a woman tried to refuse vulnerability – that is, in conventionally acceptable ways. Intense emotion, hurt, and anger, all invoked to facilitate a particular analysis of the event, were reduced to a tired reiteration of hysteria. The political was reduced to the personal, safety was (re)secured by locating the threat as external, and community was re-created through the exclusion of dissent. Exclusion here served as "the very precondition and possibility for 'community'" (Butler 1992, xiv). Any unsettling of neat boundaries between feminist and non-feminist, men and women, safe and violent, internal and external, success and failure was not welcome.

Conclusion

I suggest that the Clayoquot Peace Camp can be understood as a microcosm of eco/feminist politics, where new possibilities came into focus, people embraced an alternative life/politics and were changed forever, and eco/feminism was practised and its meanings were negotiated and contested. It also demonstrates the difficulty of sustaining transformatory politics on a daily basis. The episode discussed above is not an exception that can be ignored; nor is it evidence of the "failure" of the peace camp. No single story suffices; instead, all stories and the contestations over stories are important. As Magnusson and Shaw (2003a, 283) noted, "Precisely because Clayoquot is a hopeful sort of place, a place where the new politics (as frequently conceived) has been practiced for some time quite effectively, it may work as a reminder to us that the new politics (if that's what it is) is no easier than the old." The incident of the attack points to the crucial importance of process and transparency, and a stringent analysis of power and its unequal circulation, in the praxis of alternative politics. Because much feminist and theoretical cultural discourse, and sometimes political activism, shies away from such work, the social transformatory project of feminism remains limited. The term "essentialism" is crucial in understanding how and why this work is avoided. Rather than dismissing eco/feminist non-violent activism for being essentialist, we can examine the deployment of essentialism and trace its effects. We can also trace the damage that results from the disavowal of eco/feminist activism, both to the narratives of eco/feminism and to those whose lives constitute these narratives. The blanket application of essentialism to instances of eco/feminism negates the often complex contestations and negotiations of the process of eco/feminist politics. Naming eco/feminist politics as essentialist functions to exclude much activism from eco/feminism and ignores the way that such politics might be usefully understood as interrogating eco/feminist orthodoxies.

Reducing the events at Clayoquot to an essentialist or strategically essentialist politics would do a disservice to activists, to eco/feminist theory, and to narratives of eco/feminism. In fact, the differing accounts of the attack on Fireweed demonstrate the limitations of Sturgeon's (1997) hope that strategic essentialisms will be destabilized by radical democratic processes, as a strategic essentialism was invoked to stabilize a moment when notions of gender and eco/feminism were being called into question. Eschewing the straightforward categorization and disavowal of peace camps as maternalist and essentialist, and choosing to examine how the category of woman is invoked and to what political ends, has much to offer. Rather than focusing on whether eco/feminism or specific eco/feminist activisms, texts, or practices are essentialist, or anti-essentialist, attention shifts to examining the work in which particular mobilizations of the category of woman, or nature, or essentialism are implicated. Thus, the figures of the feminist peace activist or the feminist peace camp are not overdetermined in a way that negates the possibility of transformations in meaning; eco/feminist peace camps can be understood as sites of struggle over the meaning of woman and the practice of eco/feminist politics, where both are not just reified but are also refigured.

As Clayoquot reveals, eco/feminist non-violent direct action continued in the 1990s but was not recognizable through the espousal of a maternalist politics or a separatist, women-only peace camp. The apparent invisibility of feminism in the anti-globalization movement (see, for example, Eschle 2004) could be recast as a failure of academics to perceive that feminist activism is no longer identifiable through recognized signifiers such as women-only activism or a focus on gender issues. Eco/feminist activism in the global justice movement persists (Starhawk 2002). Furthermore, issues of violence in the anti-globalization movement (see Sullivan 2004) suggest the inadequacy of maternalism and essentialism for understanding non-violent activism and that the need for a continuing eco/feminist theorization of non-violence remains urgent. It is necessary to return to sites such as eco/feminist peace camps and *examine* rather than *assume* the essentialism purported to be rife, in order to recognize contemporary eco/feminist activism. Eco/feminist activism might then be understood as an important site for the interrogation of eco/feminist theory, demonstrating neither the simple persistence of maternalist peace politics nor the progressive emergence of anti-essentialism, but rather the messy and complicated ongoing work of politics.

6

Mothers, Grandmothers, and Other Queers in Eco/Feminist Activism

> My son, at the time, while this was going on and we were hearing about Clayoquot in the news and a lot of Denman Islanders were really involved in it, he was having a really hard time with um, the state of the world ... So he was seven. And he was a really astute guy, and he's really concerned with global stuff. He thinks about people on the other side of the planet not having a bed to sleep in, you know, and stuff and, you know, these things bother him. And he was um, he was um telling me he was afraid that his kids would have to wear gas masks when they were his age and, or when they were born; he was tearful about it, he was really frightened about it, and feeling really helpless. So it felt like Clayoquot was an opportunity for me to say, "You know, there *is* something that we can do, there's even something *I* can do – your mother can do this! I'm gonna do this, you know?" So it was just kind of, my main motivation was to try and ease his fears and have him not feel quite so helpless.
>
> – KIM BACK, AUTHOR INTERVIEW (1996)

> So anyway, we got to the camp, and that evening, yeah, I went for a walk, just up, just you know, we were just camped in the truck, in the overflow place, and just, went for a walk on my own that evening, and just seeing – just all that devastation, and ... I was trying to think when [my granddaughter] was born, she's just five now, so she was just one, and, I guess, I just, just the whole thing of just being a grandmother, and realizing my, this is my responsibility, and how important she was, and what a mess we had made of the world, and just wanting, just wanting a world for her, and so that's really ... what I did it for.
>
> – JANE FAWKES, AUTHOR INTERVIEW (1996)

The Nature of Motherhood

Motherhood and grandmotherhood appear to be central in the two accounts above. Stories in which motherhood spurs environmental activism are not unexpected. In fact, one might anticipate little else – many writers have understood women's environmental activism through the lens of maternalism.[1] Women, as mothers, have been perceived as the particular victims of environmental degradation because pollutants and toxins often manifest themselves first and most visibly in the bodies of pregnant women through miscarriages, fetal abnormalities, and birth defects, and through illnesses in children. This argument relies on the specificities of women's bodies, particularly their reproductive capacities, to make claims about their vulnerability to environmental degradation and to account for a distinctive eco/feminist politics. As well as relying on a biological account of women as mothers, explanations of women's environmental activism often invoke women's work of caring for children. Such accounts produce women's environmental activism as a politics founded on a moral economy of women as victims, or conversely, as saviours of the planet. Women as primary carers for children often come together to act collectively and politically when they realize that they and/or the children in a community have become ill due to environmental toxicity.

Yet, a politics based on the specificity of women, particularly on motherhood, has been contentious not only for many feminists who are critical of eco/feminism, but also for many eco/feminists themselves. Catriona Sandilands (1999, xiii) began her book on eco/feminism by critiquing accounts that, as she put it, "boil down to an obvious manifestation of natural protective instincts towards home and family." She (ibid.) argued that "the neoconservative aroma of this discourse should be quite noticeable: a return to patriarchal and heterosexual 'family values' will restore not only a healthy (natural) family but a healthy (natural) planet." Such a narrative served "rather uncritically as a central one for so many discussions of feminism and ecology" (ibid., xv). Although Sandilands does make an important point, critics of maternalism in eco/feminism rarely ask why any eco/feminist – that is, any feminist – would want to support such conservatism. As I have suggested with respect to the Chipko Movement, some eco/feminist academics critique maternalist arguments as a way of distancing themselves from supposedly essentialist politics and of establishing their own theoretical sophistication. Another route might be to resist such an easy reading of maternalism as retrograde and to explore what else might be going on in such accounts.

In this fraught context, the epigraphs that head this chapter are intended as provocations in pursuit of the question "How has maternalism-as-essentialism become the most obvious reading of eco/feminism, to the exclusion of any alternative possibilities?" Producing possible "evidence" of maternalism in the chapter epigraphs was in fact difficult, despite its supposed ubiquity in eco/feminism; the passages required careful editing to be sufficiently suggestive of maternalism. I invite readers to construe them in the contexts of the many hours of interviews I carried out and so to appreciate that even these short quotes already exceed the meaning that I suggest many might read into them.

It is worth emphasizing that maternalism was not generally invoked collectively, or publicly, in connection with the Clayoquot camp. For instance, although there was a women and children's day of arrest, it was "women and children," not "mothers and children." Other collective days of action were associated with business people, forest workers, seniors, doctors, and residents from various Vancouver Island communities, all of which undermines easy recourse to a simplistic dismissal of identity politics at Clayoquot. Motherhood was explicitly invoked at a few junctures. One of these, which was unplanned, occurred at the blockade during the first day of arrests. The thirteen arrestees were taken to Ucluelet jail for processing, an event that would be repeated every weekday throughout the summer. Three of them refused to sign the undertaking not to return to the logging road and so were held in jail, pending trial. These three, Betty Krawczyk, Inessa Ormond, and Judith Robinson, were all mothers and grandmothers, a fact that they themselves and others invoked while discussing their treatment. In the end, all spent more than three months in jail, awaiting trial. Below I also discuss another manifestation of grandmothers at the camp, the Raging Grannies, a non-violent direct action group.

In this chapter, I pursue my interest in challenging the reduction of maternalism to essentialism, and the need to exclude motherhood from the domain of politics. I begin by tracing maternalism's histories in the peace movement. I then outline divergent responses from eco/feminists to the supposed maternalism of eco/feminism, turning to Noël Sturgeon's work and her cautious defence of maternalism as an instance of strategic essentialism. Sturgeon remains unusual in her refusal to reject eco/feminism's apparent maternalism out of hand. Nonetheless, I suggest that this analysis does not go far enough. I also turn to theorists such as Cate Sandilands, Chris Cuomo, and Sherilyn MacGregor, who are more explicit in their rejection of maternalism, or what they term "ecomaternalism" or "motherhood environmentalism." In place of the figure of the mother, these theorists propose alternative figures for eco/feminist politics: the cyborg, the queer, the citizen. Finally, I consider motherhood and grandmotherhood in the accounts of women who were arrested at Clayoquot to produce an alternative account of maternal politics.

Mothering Nature and Feminist Peace Politics

Critiques of maternalism in eco/feminism need to be contextualized by examining what has been at stake for feminists in these debates and how motherhood has figured in narratives of feminism more broadly. Many feminists have identified motherhood as oppressive for women. For example, the late Shulamith Firestone (1970) argued that women's biological child-bearing capacities comprised the root of their oppression; she suggested that developing technology to relieve them of the burden of child bearing, thus ending the sexual division of labour, would resolve the problem. Among the initial demands of the women's movement were an end to compulsory motherhood, access to safe and reliable contraception, abortion on demand, and free nursery and childcare (Fairbarns 2003).

The late Adrienne Rich's *Of Woman Born: Motherhood as Experience and Institution* (1995) is frequently cited as shifting the focus from mothering as a source of oppression to a recognition that it is also a source of joy and creativity (O'Reilly 2004; Hallstein 2007, 2010). Rich (1995, 13, emphasis in original) distinguished between two meanings of motherhood: "the *potential relationship* of any woman to her powers of reproduction and to children; and the *institution,* which aims at ensuring that that potential – and all women – shall remain under male control." Rich's (ibid., 14, emphasis in original) specification that her book was "not an attack on the family or on mothering *except as defined by the patriarchal nuclear family*" is crucial but is sometimes lost in subsequent discussions of motherhood. It was not motherhood, or even the family per se, that was the subject of her critique or that of many feminists, but rather motherhood and the family as constituted in the hetero-patriarchal nuclear family.[2]

Although Rich did not articulate a specifically maternalist feminist politics, others, particularly in feminist peace politics, did so (see Elshtain 1981, 1982; Ruddick 1989; Elshtain and Tobias 1990). Women's peace movements that are often identified as maternalist include the Women's International League for Peace and Freedom, Voice of Women, Women's Strike for Peace, and peace camps from Greenham Common to Seneca. The late Sara Ruddick (1989) developed the concept of "maternal thinking" to describe the situated knowledge that arose from the practice of mothering, the work involved in sustaining children, which, she stated, could also be undertaken by men.[3] Ruddick argued for a conception of citizenship that would incorporate maternal thinking.

Critics have insisted that maternalist politics elevate women's identity as mothers above all other identities or relationships and that diverse experiences and practices of motherhood are negated or subsumed under a universalizing rhetoric. Some have suggested that they can ignore women's own concerns about themselves, that they perpetuate a (Christian) sacrificial logic, and that they uncritically celebrate motherhood, failing to recognize the powerlessness of mothers and relying on and reinforcing the traditional iconography of motherhood.[4] Unsurprisingly, feminists have been concerned about maternalist arguments that stem from patriarchal framings of women's responsibilities as mothers and that have often been used to justify women's exclusion from politics, education, work, and the public world. It is worth noting, however, that there is no explicitly articulated maternalist eco/feminism. Nonetheless, the terms "ecomaternalism" (MacGregor 2006) and "motherhood environmentalism" (Seager 1993; Sandilands 1999) have been used to define and critique a position not actually claimed by any eco/feminist (as is also the case for other typologies identified by Sturgeon 1997).

At the same time, many eco/feminists have been intensely critical of apparent instances of maternalism in eco/feminism.[5] Yet critics of so-called ecomaternalism necessarily ignore the evidence of extensive eco/feminist criticism of maternalist rhetoric. Joni Seager's *Earth Follies* (1993) charted the use of maternalist rhetoric by politicians, corporations, and environmentalists alike. For example, she (ibid., 221, emphasis in original) quoted a comment made by the vice-president of Exxon following the Valdez oil spill in Alaska: "I want to point out that water in the [Prince William] Sound replaces itself every 20 days. The Sound flushes itself out every 20 days. Mother Nature cleans up and does *quite* a cleaning job." The use of maternalist rhetoric "as a cloak for paternalism" is hardly unique to environmental politics (Koven and Michel 1993, 18). Catherine Roach (1991, 2003) noted that the environmental movement has employed patriarchal notions of motherhood via slogans such as "Love your Mother," which appear beneath images of the earth and in posters, bumper stickers, advertisements, and other eco-paraphernalia. Drawing on Roach's work, Catherine Belmont (2007) examined the "natural disaster heroine" in popular films, in an article that echoed the concerns of Sandilands. Implicit in these eco/feminist engagements with maternalism is an attentiveness to context, unlike other accounts, which often treat all apparent maternalisms as if they are equivalent. Interestingly, Roach (1991) suggests that "Mother Nature" and "rape of the earth" might have different meanings for men and women (if not explicitly feminists and non-feminists, or different kinds of feminists, including men). However, critics, whether feminist or eco/feminist, rarely acknowledge this specificity. For instance, Sandilands (1999) does not distinguish between invocations of motherhood by the Ontario Advisory Council on Women's Issues, "scientific" studies on women's participation in environmental politics, images of mothers in green product advertisements, or the stories of Lois Gibbs and Rachel Carson. There is a risk in treating differing accounts of motherhood as if they are all manifestations of the same thing.

Furthermore, though some eco/feminists may invoke motherhood at various points, they generally also articulate other accounts that would not necessarily be consistent with an essentialist maternalist politics. For instance, Gaard and Gruen (1993, 11) proposed that "ecofeminists claim that environmental issues are feminist issues because it is women and children who are the first to suffer the consequences of injustice and environmental degradation," a comment that could be understood as illustrating the kind of maternalism cited by Sandilands (1999). Yet this would not do justice to Gaard and Gruen's work. Elsewhere, Gaard (1998) has been critical of maternalism, and in her book *The Nature of Home* (2007), she has reflected thoughtfully on her own experience of becoming a mother.

My point is not that theorists such as Gaard are inconsistent, or that they are not really maternalist, or only accidentally so, or that they later recognized the error of their ways and either moved beyond maternalism or reverted to it. Rather, these supposed inconsistencies indicate that motherhood is not something that we can be for or against, or that we can get beyond. Instead, it is an important aspect of everyday life for many people, including many activists, and eco/feminists therefore need to find more adequate ways of theorizing it. The reference to Gaard's work also reminds us that, whereas Sturgeon turns to radical democratic activism to destabilize strategic essentialisms, theory also has the capacity to destabilize itself, and, as I suggest through my use of oral histories, that narratives and life stories also have this ability.

Hysterical Housewives in the Anti-Toxics Movement

Maternalist accounts of women's involvement in environmental activism often refer to the anti-toxics movement, particularly in the United States (Merchant 1996, 136-66).[6] Lois Gibbs (1998) is an archetypal example here. She turned to activism when her son became ill, like many other children in her neighbourhood, Love Canal, in Niagara Falls, New York. Her investigations led to the discovery that the neighbourhood was built on a toxic waste dump; following her campaigning, over eight hundred families were relocated, and a cleanup of the area began. Hers is a classic story of a woman who became an activist through the illness of her children and who campaigned on behalf of them and everyone in her neighbourhood; it combines both a biological account of women and children's victimhood and a social account of women's roles as mothers in motivating and sustaining activism.

Examining accounts of anti-toxics activism that do not come explicitly from feminists or eco/feminists is instructive because, though maternalism is often presented as motivating activism, it is also described as being transformed in the process. Sociologists Phil Brown and Faith Ferguson (1995, 148) offer a version of this analysis in their research:

> In each case study we have found, there are remarkable similarities in women activists' transformations from housewives to activists ... Each woman who becomes a toxic waste activist first suspects that there may be a health problem in her neighborhood when her children become ill. Outreach to her neighbors leads to the discovery that there is a pattern of illness in the community.

Harriet Rosenberg (1987, 180) also wrote of this "transformation of isolated housewife to activist," and Cynthia Hamilton (1990, 221) observed that it can prompt changes in the home, with husbands becoming more involved in childcare, cooking, and other household tasks: "Meetings in individual homes ultimately involved children and spouses alike – everyone worked and everyone listened. The transformation of relations continued as women spoke up at hearings and demonstrations and husbands transported children, made signs, and looked on with pride and support at public forums."[7] This trajectory from housewife to activist is often presented as evidence of feminist transformation and radicalization: previously apolitical women who develop a feminist consciousness and no longer see themselves "just" as mothers can now be claimed as political activists engaged in a mass social movement. It is worth observing, however, the recognition in this work that husbands do not always respond well to the change; there are many stories of domestic tension, conflict, and abuse in response to women's organizing.[8] So, though motherhood and being a housewife are presented as a source of activism and as a site of positive change, activism itself can mark the breakup of the nuclear family. These accounts are intriguing because, rather than understanding such activism as exhibiting maternalism or essentialism, strategic or otherwise, they identify a transformative dimension, a changing identity, related to the activism, and they focus on motherhood as experience rather than as ideology.

It is also interesting that work on maternalism in the environmental justice movement more broadly often enjoys a favourable reception by eco/feminists who are otherwise dismissive of maternalism. For instance, though Sturgeon criticizes the peace movement's maternalism, she appeared more open to motherhood in the environmental justice movement, as discussed in the work of Giovanna DiChiro (1992, 113), who stated that many women become involved because of "deep concerns about the health and future survival of their children and their communities." However, DiChiro (ibid., 115) was reluctant to place *too much* emphasis on women as mothers, writing that women's identity "as simply 'mothers' is by no means always the central focus of their activism." So, though neither DiChiro nor Sturgeon ignore mothering, arguably their analyses do not go far enough. These, and other, accounts risk opposing motherhood and feminism by asserting that if women wish to become feminists, or to be recognized as feminists, they must diminish the importance of motherhood in their lives. In this teleological narrative, women must progress from mothers to activists, leaving motherhood behind in their pre-feminist past or in a separate private/domestic space. This perspective appears to rebuke mothers whose concerns about their children lead them to activism; if they wish to gain admittance to the theoretically sophisticated club of feminists and eco/feminists, they must have better motivations than this.

Maternalism as Strategic Essentialism

Noël Sturgeon's *Ecofeminist Natures* (1997) has provided one of the most generous accounts of maternalism in eco/feminism. In the first instance, Sturgeon (ibid., 70) disputed just how pervasive such "Moral Mother imagery" has been in the non-violent direct action movement, which correlates with my account of the lack of extensive maternalism at Clayoquot. When she did identify maternalism, she defended it as strategically essentialist, arguing that such essentialisms are commonly destabilized in the context of radical democratic direct action politics. Nonetheless, Sturgeon (ibid., 59) acknowledged "the political dangers of using such symbols as 'Mother Nature,' which may reinforce patriarchal assumptions about the more 'natural' status of women." She (ibid., 69) added,

> Certainly it is the case that recurrent features of the direct action movement have been symbols and language that make uncritical connections between women, nature, and mothering: the use of Mother's Day for women's antimilitarist actions, an analysis of militarism that connects it with male violence against women and the "rape" of the earth; the prominence of feminist pagans, the witches that weave webs; and, within the movement, the frequent parallels, or even equations, made between feminism and non-violence, feminism and anti-hierarchical politics, and feminism and environmentalism.

Sturgeon's approach was to examine movements for debates regarding the meanings of "women," which could act to destabilize essentialist tendencies. For instance, she discussed the 1987 Mother's Day Action at the Nevada Test Site. Concerned that women who were not mothers would feel excluded from its campaign, the local Santa Cruz group, with which Sturgeon herself was involved, suggested a Mothers and Others' Day Action. The national organizing group welcomed this idea, but it interpreted "Mothers and Others" to mean women and men, thus conflating women and mothers, and suggesting anxieties about different exclusions. Sturgeon used this example to demonstrate that, just as movements can destabilize essentialisms, they can also restabilize them. Some men who attended the action wore pillows under their T-shirts, a gesture that Sturgeon (1997, 72) understood as "pregnant with meaning," as the men claimed their status as mothers, illustrating autonomy within collective actions and constantly challenging any fixed notions of what feminism or women should be. Sturgeon (ibid., 73) argued that the "oppositional consciousness of the movement deployed various images of 'mothers' (men as mothers, grandmothers, lesbian mothers, Mother Earth, etc) in an attempt to replace the Reaganesque image of a militarized white mother of two who willingly supports the sacrifice of 'her' men and children in war and in nuclear industries."

Sturgeon (ibid., 74) addressed the possible pitfalls of invoking strategic essentialisms through posing the question "But what about the danger of women becoming reinscribed into patriarchal discourse?" As she (ibid., 75) explained, this was not a problem, since "the mainstream press and the people who lived in the communities in which these actions have taken place rarely perceived women in the movement as 'traditional' wives, daughters and mothers." Commonly, it was women-only, or women-led actions that attracted the most aggressive responses, with Sturgeon noting that "gay-baiting is a common reaction to women's actions." Significantly, Sturgeon focused solely on the possibility that "ordinary" people might reinscribe women in patriarchal discourse, rather than asking whether eco/feminist academics might do the same. Whereas Sturgeon (ibid.) read it as "ironic that where feminist observers see dangerous essentialists, patriarchal observers see dangerous lesbians," the real irony here is that the local communities and the mainstream press (presumably non-feminist or even anti-feminist) would have been perfectly capable of reading such activism as feminist, as posing a threat to hetero-normativity, to conventional notions of mothering, families, and community – as an unnatural challenge to cherished family values. On the contrary, some feminists appear to see quite the opposite, the consolidation of a particular version of mothering – quint-essentialists – not "unnatural women" at all, but rather the most natural of all women, the mother. This, I think, is an intriguing gap in readings.

Cyborgs, Queers, and Citizens against Mother Nature

The figure of the mother in eco/feminist activism and theory has sometimes been displaced by alternatives – the cyborg, the citizen, the queer, and the practice of drag. For Sandilands (1997, 18),

> Judith Butler's work on drag and Donna Haraway's on cyborgs inspire a rather more interesting series of political metaphors than the Mother – or sister – Nature who has tended to inhabit most ecofeminist writings. Queers and cyborgs are not easily gendered or natured, and thus represent a new kind of character to inhabit the shifts and fissures of identities in collision and collusion.

Sherilyn MacGregor (2006, 75) suggested that "the production of an open-ended and expressly political identity – like citizen – would be a better move for ecofeminism than the production of a narrow, pre-political and overdetermined identity like earthcarer or mother environmentalist." However, I understand eco/feminism as a movement that questions hierarchies and dualisms, so I want to refuse this opposition of mothers with cyborgs, queers, and citizens. I think it possible, even necessary, to read Mother Nature as a queer, or a cyborg, or even a queer cyborg, though I am less sure of her citizenly roles. Mother Nature is surely a queer coupling – a cyborg even, a creature of breached boundaries.

At the same time, the figurations of queer, drag, the cyborg, and the citizen have all proved more open to domestication than their originators had anticipated. After all, Butler and Haraway have had to clarify their intentions, following the wildly unexpected success of their creations. Butler (1993, 230, emphasis in original) pointed out that she intended drag as an *example* of performativity, whereas some took it "to be *exemplary* of performativity." Haraway's own "going to the dogs" suggests that the cyborg has proven only too potent a metaphor for her own intentions. As she wrote in *The Companion Species Manifesto* (2003, 4), "The present manifesto asks which of the two cobbled together figures – cyborgs and companion species – might more fruitfully inform livable politics and ontologies in current life worlds." Though she (ibid., 4-5) conceded that "these figures are hardly polar opposites," she also distinguished them:

> I appropriated cyborgs to do feminist work in Reagan's Star Wars times of the mid-1980s. By the end of the millennium, cyborgs could no longer do the work of a proper herding dog to gather up the threads needed for critical inquiry. So I go happily to the dogs to explore the birth of the kennel to help craft tools for science studies and feminist theory in the present time, when secondary Bushes threaten to replace the old growth of more livable nature-cultures in the carbon budget politics of all water-based life on earth.

With the figure of the dog, Haraway seems more explicit in her intentions. She (ibid., 5) wrote, somewhat didactically, as if trying to stave off the inevitable appropriations of her concept, that "dogs are not an alibi for other themes; dogs are fleshy material-semiotic presences in the body of technoscience. Dogs are not just surrogates for theory; they are not just here to think with. They are here to live with." Perhaps in the end the cyborg has proved as susceptible to co-option as the figure of the mother or the goddess, albeit in different ways, ultimately as overdetermined as Mother Nature herself.

These re-visions remind us that such figures were not intended as universally or timelessly subversive. Though we might easily assume that companion species are simply a better and more up-to-date figure than cyborgs, they are more clearly "situated knowledges," creatures of different times, places, and intentions, with different work to do. Yet though Butler and Haraway appear clear that their re-visions do not constitute progress narratives, they have often been invoked in this way by others. Butler (1990, 148-49) insisted on an understanding of identity politics as a mode of representation, as a way of accounting for politics and not as a fixed historical moment:

> The deconstruction of identity is not the deconstruction of politics; rather it establishes as political the very terms through which identity is articulated. The task here is not to celebrate each and every new possibility *qua* possibility but to redescribe those possibilities that already exist, but exist within cultural domains designated as culturally unintelligible and impossible. If identities were no longer understood as part of a set of practices derived from the alleged interests that belong to a set of ready-made subjects, a new configuration of politics would surely emerge from the ruins of the old.

Butler here usefully explains the limits of the cyborg, the citizen, and the queer in drag, as well as the work that is required for politics. For instance, though Sandilands draws on a performative approach to the cyborg and of the queer, she did not develop a performative approach to motherhood. She (1997, 8) called for "a project in which the lives and desires and bodies of actual people – queers and cyborgs – sit front and centre in ecofeminism's discursive challenges." This risks implying that mothers are not "actual people" and that mothers and queers are mutually exclusive categories, which Sandilands undoubtedly would not have intended.

Whereas the alternative figures of the queer, the cyborg, and the citizen are offered as if they pose resolutions to the dilemma of maternalism, of women and their uncertain natures, they seem only to displace the supposed problem. Jana Sawicki's (1991, 13) account of how feminism might engage with motherhood seems useful for eco/feminists: "A Foucauldian feminist could also embrace mothering theory to the extent that it was historically and culturally specific and did not claim to be the definitive account of the origins of male domination." Sawicki added (ibid., 60), "Without rejecting mothering theory, the genealogist adopts a critical attitude toward it ... Rather than assume that 'mothering' represents a unitary phenomenon, the genealogist looks for discontinuities between practices and ideologies, and between practices across cultures." This approach seems to offer fruitful possibilities for eco/feminists who want to think about how motherhood can be part of the story of eco/feminist theory and activism.

I turn next to stories from lesbian and heterosexual mothers and (great) grandmothers from the Clayoquot campaign to illustrate how activists are challenging conventional practices of mothering and the reproduction of the heterosexual, patriarchal family that is at the heart of many feminists' concerns over motherhood. I turn to women's narratives to demonstrate that they are actively engaged in the project of reconceiving motherhood. For me, the Raging Grannies provide a useful way to begin to redescribe the lives of women environmental activists who are mothers and grandmothers. I ask whether the Raging Grannies, and later Kim Back and Jane Fawkes, are best understood as rabid essentialists, or cyborg creatures, or "old women who have gone to the dogs." Which descriptions might best serve eco/feminist politics? Beginning with the performative politics of the Raging Grannies, with their parodic granny drag, I then move on to discuss what might appear to be more mundane accounts of motherhood in the narratives of Clayoquot activists.

Granny Drag: The Raging Grannies

A NEW TEDDY BEARS' PICNIC

If you go down in the woods today
You'd better not hug a tree
If you sit down upon the ground
They'll call in the RCMP

> Protecting them is a terrible crime
> Destroying them is perfectly fine
> 'Cos that's the way the NDP runs this province.
>
> – JEAN McLAREN, *SPIRITS RISING (1994)*

If you go down in the woods today in British Columbia, or many other parts of Canada, along with the clear-cutting of trees, you might just find the Raging Grannies protesting, or running a song-writing workshop for children.[9] Or you might encounter them in the city, singing a new version of "Teddy Bears' Picnic" in the foyer of a multinational corporation. The Raging Grannies were founded in Victoria, on Vancouver Island, during the late 1980s but now exist in many countries, including the United States, Germany, and Australia. Greenpeace is not the only global direct action group to originate in British Columbia. The Grannies use street theatre, and especially songs and dressing up as archetypal grannies, to protest a range of issues including deforestation, poverty, nuclear war, militarization, and domestic violence, to name just a few. Jean McLaren, commonly invoked grandmother of the Clayoquot Peace Camp, was just one of the many Raging Grannies who were there. A member of the Gabriola Island Grannies, Jean lived at the camp all summer long, running some of its non-violence training workshops and subsequently writing an account of it in *Spirits Rising* (1994). The Victoria and Saltspring Island Raging Grannies visited the camp on a number of occasions (ibid., 57), Grannies came to be arrested on the Seniors' Day of Action, and many came to sing and to be arrested throughout the summer. Although I interviewed Grannies such as Jean McLaren, in what follows I also draw on *The Raging Grannies Songbook* (McLaren and Brown 1993), *Off Our Rockers and into Trouble* (Acker and Brightwell 2004), and *The Raging Grannies* (Roy 2004).

Being a Raging Granny is not about babysitting or organizing wholesome picnics in the woods with anthropomorphized teddy bears. Nor is it about protecting children from the potential dangers of the world. Rather, being a Raging Granny is about enabling children and other Grannies to learn about the world, about how to rage and take action, and how to have fun in the process. With all their raging, the Grannies have precious little time for babysitting; voicing a noisy concern and fury about the present and the future, they oppose quiet, sentimentalized, and nostalgic images of grannies that are wedded to the past. Through their activism, the Grannies have claimed access to places where grandmothers are not expected to go. After all, grandmothers are not supposed to be very mobile; they should be confined to the home, preferably the fireside, telling suitably moral fairy tales to grandchildren who have heard them many times before. Yet the Raging Grannies are at home in the woods and in the foyers of huge corporations. They rage about the state of the world, as McLaren and Brown (1993, 4, emphasis in original) explained:

> Why *Raging* Grannies? What are we raging about? We are enraged about the state of the Earth we are leaving for our precious grandchildren. We are raging against the system that has allowed this to happen and the institutions that perpetuate the atrocities against our planet. As grandparents we have a responsibility to the children and grandchildren of the future, and it is not too late for us to act.

The Raging Grannies do not make martyrs of themselves for the sake of their children, grandchildren, the planet, or the future. There is no sacrificial logic; they do not adopt a position of moral superiority and are not moral mothers or grandmothers. They act not from a position of innocence, or even of victimhood, but from that of responsibility and accountability, for children, for the planet, and implicitly and sometimes explicitly, for themselves and other grannies. Many came of age during the post-war era of unthinking consumerism. Unlike many of the politicians and corporations they challenge, the Grannies do not disavow their responsibilities in the world; after all, some of them voted for the NDP government that permitted logging in Clayoquot and are only too aware of matters of accountability. In fact, some recognize that their own post-war lifestyles have contributed to the environmental crisis, which intensifies their need to become involved and to take action. The Grannies' aims are high. Instead of hoping to accumulate financial wealth or a house to bequeath to their grandchildren, they want to leave the whole planet. And if a whole, restored planet cannot be guaranteed, at least they will pass on skills about treading lightly, taking action, and having fun:

> As Grannies, we aim to be an inspiration to older women to speak up about issues of justice and peace and the environment; to be interested in public affairs and in issues affecting the population as a whole. Ideally, we should use our granny status not only to change the world, but to change our peers – to have a lot of fun, and to empower older people to do

> something for the future of the Earth so there is something to pass on to our grandchildren. (McLaren and Brown 1993, 22)

Importantly, being a Raging Granny is not defined biologically through having children or grandchildren. *The Raging Grannies Songbook* reassures readers that "being a real grandmother is not even necessary – you only need to have what we call 'Raging Granny mentality'" (ibid., 23). It is even possible for children to be Raging Grannies; the Grannies run workshops on how to be a Granny, understanding the practice of being a Granny as learned and shared rather than innate. Furthermore, the Grannies are not founded on a common identity or the expected interests of grandmothers; rather, they are premised on challenging the granny stereotype, through parodying popular ideas about grandmothers. As McLaren and Brown (ibid., 21) put it,

> The unique feature of the Raging Grannies, however, is that we actually take the granny stereotype and use it to do our political work ... Raging Grannies gain entry to forbidden places – for who would deny access to sweet little old ladies? We take advantage of the stereotype – play up the image for all it's worth – to get into inaccessible political offices and deliver our message. Using songs sung to familiar old tunes adds to our impact. The Grannies are able to get people listening before they know what they are actually hearing, so that songs are often half over before security guards and others in charge realize what is really happening. Through astute and wily use of the stereotype, the Raging Grannies catch the authorities unaware and get heard far and wide.

The Raging Grannies are best understood as parodying the notion of grandmothers, as calling the identity into question, rather than as manifesting any essentialism, strategic or otherwise. Singing re-visioned versions of familiar songs, they make extensive use of humour, irony, and comedy. Writing on parody, Butler (1993, 138) argued that "the notion of gender parody defended here does not assume that there is an original [grandmother] which such parodic identities imitate. Indeed, the parody is of the very notion that there is an original ... So gender parody reveals that the original identity after which gender fashions itself is an imitation without an origin." Thus, being a grandmother is revealed as an effect through the Grannies' use of what might be called granny drag. Through their theatrical renditions of granny-ness, their parody of granny stereotypes, they locate "grandmotherness" as an effect, as produced and unnatural, rather than natural:

> Closely connected to this is the tradition of clowning that the Raging Grannies carry on. By dressing up in grandmotherly dresses, shawls and flowery hats, and doing our routines, many Grannies feel the power that clowns possess to say things that they otherwise could not. "The costume changes my whole personality," as one Granny puts it, and "I'm no longer afraid to get up and do something about the world. There are a lot of women who'd like to do what we are doing, but don't have the nerve or confidence." (McLaren and Brown 1993, 21)

But the Grannies don't actually own such grandmotherly clothes themselves and must find them in second-hand shops. They combine flowery dresses, shawls, and antique brollies with saucy hats, sequins, garters, and fluorescent pink shoes – the desired effect is outrageous rather than endearing, a clash of sweet and tarty that underscores the supposed asexuality of older women, an image that is also challenged in their "Safe Sex" song:

> Hey Ho! Hey Ho! As off to bed we go
> We grannies smile because we know
> Safe sex is quite the best. (ibid., 55)

Butler (1993, 241) described performativity as "a relation of being implicated in that which one opposes, [a] turning of power against itself to produce alternative modalities of power, to establish a kind of political contestation that is not a 'pure' opposition, a 'transcendence' of contemporary relations of power, but a difficult labor of forging a future from resources inevitably impure." Whereas Butler used drag as an instance of gender performativity, the Raging Grannies dress up as *old* women. Thus, they perform not only gender and sexuality but the imbrications of age and gender, not just any femininity but an ageing femininity – a rather queer transgenerational feminist activism. Recalling Butler's (1990, 146) argument that "bodily surfaces are enacted as the natural" and that "these surfaces can become the site of a dissonant and denaturalized performance that reveals the performative status of the natural itself," the Grannies' street theatre refutes any account of grandmotherhood and ageing as the end of womanhood, as a descent into a supposed asexuality and a return to nature through decay and death. Although the Grannies may appear to be implicated in what Lee Edelman (2004) calls a "reproductive futurism," their activism cannot be reduced to the use of the sentimentalized child to sustain our collective future. Although grandmothers might seem to be a kind of übermother, doubly mothers even, mother and grandmother, it is worth remembering that that are not supposed to be sexual or to reproduce. Nonetheless, the Raging Grannies are incredibly generative – producing other Grannies of all ages through a kind of parthenogenesis, with gaggles of Grannies involved in a public orgy of singing and street theatre. Though supposedly post-menopausal, they are regenerating in many parts of the globe and are challenging conventional practices of mothering and the reproduction of the heterosexual patriarchal family that is at the heart of many feminists' anxiety over motherhood. For some, the activism of grandmothers may appear to embrace a sacrificial logic, to be wedded to a teleological reproductive future. Yet the Grannies' ability to discover a new lease on life disrupts the supposedly inexorable telos of death. Indeed, the Grannies appear to be having the time of their lives.

But what, then, of the possibly more mundane performativities of ordinary mothers and grandmothers, such as Clayoquot activists Kim Back and Jane Fawkes? Butler's work remains useful here. She (1994, 33) insisted that performativity be understood through the notion of resignification: "I'm still thinking about subversive repetition, but in place of something like parody I would now emphasize the complex ways in which resignification works in political discourse." I would add that political discourse can include women's oral histories and their own accounts of their experiences.

Queering Motherhood: Kim Back

When I interviewed Kim Back in 1996, she was in her thirties, mothering three children and supporting herself by doing manual work. She lived on Denman Island, which has a population of about fifteen hundred and is situated off the east coast of Vancouver Island. Many Denman Islanders were involved in various ways with the Clayoquot campaign, including a group that came to the camp in late summer to be arrested.

Motherhood was central to Kim's story of her involvement in Clayoquot – although not as usually articulated. She explained her decision to get arrested in terms of her young *son's* concern about the future of the planet and of his own children, rather than because of *her* concern for his future.

Bearing in mind Ruddick's recognition of men's capacities to engage in maternal thinking, this is suggestive of the capacity of young boys to also engage in maternal thinking – a maternal concern not envisioned in more mainstream accounts. Kim appeared to oppose motherhood and activism – "I'm like, not [a] radical ecofeminist person, you know, I'm very kind of community-oriented, and raising my kids" – but at another point she contrasted herself with her partner:

> My partner at the time, she was kind of more gung-ho about it at the time than I was, you know ... She brought up the – okay, this is happening, she was very, she likes to do the radical thing, too, I mean it's kind of her lifestyle. And I'm like – a mom. Right? I'm kind of like – I'll raise children.

Kim invoked her own experiences of mothering in explaining her decision to become involved, an explanation that would appear to be particular to her. Her partner, who was also a mother, had a long history of involvement in activism and had been arrested during an action in Alberta. Thus, Kim's account of mothering focused on her own practices of mothering rather than on a universal or general definition of what a mother is. Although she reiterated conventions of mothering, such as caring for children and being community oriented, Kim nonetheless behaved in ways that challenged maternal stereotypes. She was separated from her husband; she mothered three children and supported both herself and them by being a general "odd-job woman" on Denman; she dressed in boots and overalls, and with her crew-cut, might just be mistaken for the odd-job man. When I came to Denman, she had just organized a drag ball at the local community hall, which was attended by nearly a hundred people (a good turnout in such a small place). I interviewed her outside the Denman General Store, to constant calls of "great party" from passersby.

Although Kim presented her Clayoquot involvement as stemming from her son's concern about the planet, he had probably learned some of his ideas about environmental destruction from his extended family, including others on the island. Furthermore, she also presented the plan to go to Clayoquot initially as her partner's. I mention this not to suggest that Kim's account is contradictory, but to highlight the many factors that were involved in her decision to be arrested. Whatever its motivation, acting on the decision required extensive support and "maternal" work, as Kim explained:

> You know the plan at the time was for me to go to jail and ... we had lots of conversations about that, the kids and I – yeah. We had a family conference ... So. But we'd had enough of a conversation that we [Kim and her partner] were both feeling really strongly that we should get involved, so it was decided that I would be the one to be arrested, and she would be my support network around that, you know, the me being away from home thing, especially – the kids were, you know, they had concerns and asked questions, you know, like – "Are you going to go to jail, and what's going to happen to us?" and, but you know, [my partner] was going to be there, and they liked her ... Yup – the plan was for us all to go together and [my partner] would be there and stuff, but our relationship kinda took a nose-dive between you know, the decision making and the actual event ... Well, I didn't change the plan of going and being arrested, that already felt like it – it had become – it had begun as kind of [my partner's] idea, you know, like this was something we could do, but then it kind of, inside me, like this was something I *have* to do. So – reshuffle all the realities, you know, find childcare, and off I went.

Kim also stressed the importance of friends on Denman, who told her about the Clayoquot campaign and then supported her throughout her involvement with it. Her support person during her arrest and trial was an old friend: "We've been friends for a really long time. And I knew that she was interested in the issue, she had come to the meetings, she didn't want to be arrested but she was supportive." Additionally, for someone on a low income, such as Kim, this support was crucial:

> I think that I may not have done it otherwise, I think that it would have been – too much of a stretch. For me. For one thing, I'm like, I'm really poor. Even to – I don't own a car, for me to get out there would have been, you know ... And with the community support, there was – there was more than just the ride out there, there was like – the community was way behind us, really really behind us, so like the childcare thing happened, and then all of the support around like, travelling back and forth to Victoria for trials, we were really supported then, there was money raised to do that ... Yeah, I don't think I could have done it under any other circumstances. And, emotionally, that was good for me too. Far less fearful. I'd never been arrested. I'd never even had a ticket! You know, I never jumped the stile in the subway, or anything like that, you know! I am a fine upstanding citizen.

> So it was like trying to – kind of scary, and it was good to have all my friends and neighbours around me.

Thus, activism shifted from something done by other people to something that Kim herself took on in response to her son's anxiety, and finally to something that she simply *had* to do. It may have been sparked by her son's concerns, her partner's enthusiasm, and the support of friends, but embedded in her account was also a sense of a transformation in how she viewed herself.

Kim's version of motherhood might not have been identifiable as such to others, or deemed as acceptable as practices of "good motherhood." She saw herself as a "fine upstanding citizen," a community-oriented person who was just raising her children, but others might have seen her as a poor lesbian mother. Being a visible lesbian in a small place may prompt the intense need to represent oneself as a good mother, given that lesbian mothers are often viewed as bad or unnatural. Denman may seem idyllic, but some residents were not supportive of the way Kim chose to live, and homophobia still flourished. Kim concluded the interview by stating that the process of arrest, sentencing, and electronic monitoring had been draining, and she mentioned that she was reluctant to engage in similar activism again, at least in the near future. At the same time, however, she said that she was thinking of organizing a Denman branch of the Lesbian Avengers, a radical direct action group.

Kim's queer mothering demonstrates the limits of arguments that focusing on women's non-motherhood roles and identities will produce better theory. Such an approach entails a process of addition, or rather its opposite – a subtraction – as if motherhood can simply be excised from a woman's material life circumstances.[10]

Maternalism and Reproductive Futurity: Jane Fawkes

When I interviewed Jane Fawkes, she was in her sixties. Like Kim, she lived on Denman Island, though she had moved there after, and because of, her involvement in Clayoquot. She lived with Patrick, her husband of over thirty years, worked as an artist, and was also involved in a permaculture project with other households on the island. The mother of three grown sons, she was also a grandmother.

Jane's narrative demonstrated a reflexivity regarding motherhood. When I asked her if she saw herself as a feminist, she replied by considering her own mothering, understanding that her encounters with feminism were bound up with her life experiences. Initially, she did not see herself as a feminist:

> I really enjoyed being a mother and Patrick didn't enjoy his job very much. Actually, I thought I had the better deal. I thought I was lucky, because it was just before, because I had my children, starting in the mid-sixties, and there was, the mothers that came after me, even five or six years after, they were starting to get the feminist message that women were hard done by, that it was rotten being stuck in the house, have kids and things, and I think it did them a real disservice, because I saw them being quite, being really dissatisfied, and I saw marriages breaking up because of it, and for me, it was just a great time of my life, ten years of being into looking after children, but it was a wonderful time, I came out of it not feeling great, not knowing where I was or anything, but while I was in it, it was one of the happiest times of my life.

Here Jane recounted her earlier rejection of certain feminist discourses of motherhood, based on her own experiences of mothering. Thus, rather than a version of maternalism that might position her as agent of her life and of her granddaughter's, a story of sacrificial logic that would pit her against feminism, her narrative demonstrates an ongoing reflexivity over her experience, including that of motherhood and her relationship with feminism. In fact, this is the kind of reflexive account that would be recognizable to many anti-essentialist/post-structuralist feminist theorists.

Jane's understanding of motherhood and feminism continued to shift over time:

> I really sort of got it [feminism], that, and I'm still trying to teach my husband, he doesn't really get it yet, hard for men to see it, because they feel guilty and defensive and don't really see where they fit in. He helped make the goddess [a sculpture in Jane's kitchen, where we were sitting]. I think, well, he was afraid of losing me, so he's come a long way and done a lot of work on himself too, and that's pretty amazing because a lot of the women I meet, who've actually, y'know, gone through all their rage and abuse stuff, y'know, usually the way they've chosen is to be lesbian or just to not try and live with a man, or to get a new man or something, and I think just because, we were together thirty years and had a strong family too, it wasn't something I could chuck very easily, he's, because he has showed willingness to

> do things himself, and it's been, a whole other thing. I see myself as a feminist, but I also see, I'm not anti-men or, I couldn't be, I had three sons, and I think they are victims too, it's the system ... I guess the first five years I was sort of dealing with my own thing and probably the last couple have been living with a man, learning to love a man. That's brave, and to begin with, I could only do it with total rage and screaming and throwing things, and now I can be strong but in a loving way. It's sort of a hard one.

However, in foregrounding motherhood in Jane's account, I risk doing an injustice to her story. As I mentioned above, I chose this chapter's epigraphs, one of which comes from my interview with Jane, with the intent of making maternalism manifest. Jane herself did not begin her interview with the topic of motherhood. Like Kim, she discussed numerous subjects, and there were other, related motivations and meanings for her activism.

Jane began the interview with her decision to be arrested at Clayoquot, putting it in the context of her life at the time and her efforts to deal with childhood sexual abuse:

> Because, actually, ah, it was, yeah, for me it was just totally tied in with where I was in life ... that I was working on my own early sexual abuse ... so it was a period when I was really aware of my powerlessness with the male system and all that stuff. And then, the summer of – I knew that Clayoquot was going on, but I didn't, I was too much in my own stuff, I was just too involved in my own stuff, I didn't really have any energy or anything outside of that, you know, just dealing with it. Then, it was my husband who went and took a workshop with Joanna Macy up at Hollyhock ... And so, and Patrick has always felt sort of frustrated that he was not actually doing anything, sort of, he'd get all the magazines, so when Clayoquot happened, it was like, okay, this is what he wanted to do, so we were going to go over there. We both knew we, we really wanted to go over there, but without any thought of being arrested ... And then we heard that the local group was going, and [Denman Islander] was organizing this, and so Patrick was, I knew he was really wanting to get arrested, but I was, I just went into complete terror, I just, because I was dealing with a lot of fear stuff, so, and just, not even myself, the fact that he would be arrested, and I just didn't think I could deal with it, and so, he was ... We talked about it, and he knew how I felt, he was really wanting to but really being considerate of me, or he wouldn't have done it, if I had really, not, if I had been totally freaked out by it. So, and then, just before, we had a couple of

> meetings with the group and people came and talked about what was happening at the camp, and, and I really liked, just the people, that were involved. There was the group that had been, that had blockaded up at Strathcona, they were going to put a mine into Cream Lake, and so that was – we had come to Comox Valley at 1990, and of the people we had met and really liked, these were the people that had been involved with that and also how empowering it was, and what a difference it had made in their lives, what a fantastic experience it was, these were the same people that again were being active at Clayoquot. So then we went to a couple of these meetings, and I, I, was, I got to the point where okay, I am alright with Patrick doing this, and I told him this, this was, I'd support him doing it, and, you know, I'd go there with him and everything, and then, I thought, I'd rationalize this, this would be great, it would be really good for Patrick to do something, usually I'm the leader, really great for him to do this on his own, and that I'd support him, and it would be, I was feeling good about the whole thing, I wasn't feeling so fearful about it. And then, it was, I got out, I've all of Starhawk's books, and at that time, I really wasn't reading them, it really wasn't something, it wasn't a focus in my life, but I, I don't know, for some reason, I'd got this one, *Truth or Dare* ... and I just picked up this book and just opened it, and it was all about how women needed to, eh, eh – do something, something that was, with political action, and I just thought, "Oh shit!" it was – it just seemed like a message, and then I thought well, I thought well, I just got it that I had to do this, that, and then I thought, well, it's, you know, I'd been doing all of this other stuff, and just dealing with the fear, so it's just a question of dealing with the fear, you just do it, and so I'd pretty well decided, that was only a day or so before we went, so I had already decided that, you know, I would get arrested.

In turning from Jane's reflections on motherhood to her account of childhood abuse, I do not intend to suggest that this then is the real reason for her activism. In presenting Jane's apparent "beginning" in the middle of my account (of her account), I am not trying to set up a conservative emplotment, an answer to narrative desire. Nor am I trying to present the reader with a beginning that is now an end. At the very least, there are multiple entangled stories of what is bound up in Jane's activism and of how she herself is bound up in her own stories.

And I also want to reflect on how Jane opened the interview. Having listened (that is, having had to listen) to my explanation of why I was interested in oral histories, in something more than the moment of arrest, and of why she could talk about anything that was relevant to her, she seemed almost quietly impatient, gently tolerant. She knew at least as well as I did, if not more, what I was asking and why. So, I also do not want to mistake Jane's story for one of transformation from victim to survivor, from passivity to agency; nor do I wish to present her as the hero of her own life. This is not what I understood to be going on, not what I understood Jane to be telling me. She began in the middle of her life story, with what was present for her, which was the co-existence of different times – the continuous (but not always the same) experience of sexual abuse, motherhood, grandmotherhood, and the camp and being arrested. Her account cannot be folded back into a linear chronological narrative; nor does it separate motherhood from the other aspects of her life or isolate one moment of motherhood from the rest.

What I want to emphasize, then, is that the experience of the campaign and of the subsequent court proceedings made possible a different (relationship with) nature/self. In the following chapter, I focus on accounts of childhood natures and the romanticization of childhood. To prefigure the arguments there, Jane's story did not involve any simplistic or privileged romantic turn to nature. She told me that she "had a very lonely childhood, not being able to relate to people, so, for me sort of nature, and being alone with nature was really important." As a child, Jane found nature to be a place where she could escape the threats of home and family life, but this once safe place was now devastated. Jane's eco/feminist politics represent a complex account of herself, of (re)subjectification. The Clayoquot campaign enabled her to connect with nature in a way that was not about isolation, alienation, and disempowerment. Through challenging the authority of the logging industry, the federal government, and the courts, she transformed her experience of abuse. We might understand Jane's decision to blockade the logging road and to risk arrest as the materialization of a very literal "crossroads," to use Butler's term (following Gloria Anzaldua) – a juncture of discursive demands, a crossroads of cultural and political discursive forces. Moreover, this time she was not alone; like Kim, she stressed the importance of being supported by hundreds of others who had also gathered to protest the destruction.

I do not see Jane's story as an account of displacement, of saving her granddaughter when she was unable to save herself, or saving the planet if not herself. It is neither a retreat into motherhood nor a story of political consciousness raising from motherhood to feminism. If anything, it is a refusal of isolating these stories in space or time; all co-exist and are revisited. I understand her to say that one of the violences of sexual abuse was her sense of loneliness, in nature and in the world, of the cutting of boundaries that separated her from family and the domestic, and from the social/public world – that it carved up space, redrawing a range of relationships and boundaries. As a small girl, the only remaining space that she could find and carve out for herself was a lonely space in nature. So I understand her account, then, to be about a transformed relationship between herself and the world, or more precisely, a changed relationship with herself. Where nature and the world were once external or externalized, she is now in the world, and the world is in her.

Long before she went to Clayoquot, Jane had begun to reconfigure her family. She identified the early 1970s as the point when she and Patrick began to transform their family life. Before that time, she said, they had "been very traditional in the way we lived." Then Patrick gave up the job that he disliked and completed a master's degree. Living as students on a university campus radicalized them. Though Jane had dealt with her childhood abuse from her late fifties onward, further changes occurred. These included her feminism and subsequent reworkings of family and home life. In the aftermath of the Clayoquot campaign, she made more changes in her home and family life. Only two years previously, she and Patrick had moved from Vancouver to the east coast of Vancouver Island, but they decided to move again, this time to Denman Island to be with many of the people they had met via the campaign and who had supported them through the lengthy court proceedings.[11] They share their Denman property with other households, and though its old growth has been clear-cut, it does have some second growth. The small group of households are engaging in permaculture and trying to restore some of the forest, hoping to log sustainably at some point and to develop a small forest products business. In the meantime, they grow fruit and vegetables, and Jane continues with her artwork.

In re-examining the account from Jane that opened this chapter, we could read it not as an account of a heroic attempt to save the world for her granddaughter or of giving up personal freedom to secure her granddaughter's, but rather as a story of how Jane was moved, how she was done and undone by her encounter with the devastation at the camp, which was set in a clear-cut – done and undone by her granddaughter, done and undone by and over time. Clearly, she could not have done it all on her own. I echo Butler (2004, 19): "We're undone by each other. And if we're not, we're missing something." This "other" might also be the land, nature; and the undoing refigures boundaries and selves differently. Jane was certainly done and undone by the campaign. After her experiences at the camp, she moved to Denman, continued the process of transforming her family, shared land with five other families, healed the land, and healed herself. If there was a shift, through turn and return, it was from nature out there to a transformed notion of self.

Conclusion

Certain versions of maternalism are deeply problematic for feminists and eco/feminists, as Greta Gaard's (1993a) comprehensive overview reveals. At the same time, I am interested in an eco/feminist politics that does not rely on excluding mothers – mothers who may, or may not, also be cyborgs, queers, and citizens. Butler argued the need to maintain "women" as a site of possible resignification. In order to do this, it is also necessary to maintain the category of *mother* as a site of possible resignification. Opening up a term such as "motherhood" to "reusage and redeployment that had previously not been authorized" is one aspect of reconceiving "women." In doing so, feminists can "expand the possibilities of what it means to be a woman and in this sense to condition and enable an enhanced sense of agency" (Butler 1992, 16). The concept of motherhood needs to be reconceived as part of *any* feminist politics; otherwise, it will remain available as a discourse that can return to haunt eco/feminism. Eco/feminist politics are arguably much better served not by establishing theory or politics at the expense of mothers, but by working harder to understand the lives of mothers, grandmothers, and other queers.

7

Romanticizing the (Gendered) Nature of Childhood?

It had been absolutely necessary for truth's sake to include a few pages on our home life which for me had not been happy after the death of our parents. I had to show what drove me out to the woods and to the creatures for comfort, what caused the real starting point of my turn to Art. My family had never been in sympathy with my painting, nor entered into my life as an artist. My home life was always a thing entirely apart from my art life.

– EMILY CARR, *GROWING PAINS: AN AUTOBIOGRAPHY* (1946)

These few references to her father, from whom there never seem to be any letters or none that were preserved, betray a bitter relationship. no doubt overbearing, a clergyman with absolute authority as to the real (at least in his eyes), he must have been appalled at the thought of his daughter leaving him for the wilds of Canada. how did she manage to go against his express disapproval? did he cut her off, disallow her? ... whose truth, Ina? The truth is (your truth, my truth, if you would admit it) incest is always present.

– DAPHNE MARLATT, *ANA HISTORIC* (1990)

> In this province
> clear-cuts hide
> beyond mountains, off
> the highway and away from town.
> A family's dirty secret shoved
> into a cupboard under the stairs.
>
> – SUE WHEELER, "HOW FAR IT RUNS" (1994)

From Grandmotherhood and Motherhood to Childhood

While the overdetermination of motherhood continues to dominate understandings of eco/feminism, this chapter addresses a subject that has received relatively little attention – children and childhood. The ways in which the women I interviewed discussed their own childhoods first drew my attention to this absence of children and childhood in eco/feminist literature. Although my focus is mainly on women and their memories of childhood, I also touch on the presence of children at the peace camp.

To the extent that eco/feminist literature does refer to children, they usually appear as a source of maternal concern about the effects of pollutants and environmental degradation. Thus children are rendered as potential victims of environmental threats who need maternal protection, and even maternal(ist) activism, because they are unable to act for themselves, except in their capacity to act on their own mothers. I suggest that the lack of theorization of children and childhood presents a more potent limitation to eco/feminism than the usual critiques of maternalism.

Yet, as I have noted throughout, problems in eco/feminism often reflect broader challenges for feminism, but magnified, intensified by eco/feminism's tangled relationship with nature. Thus, feminism too has had a rather limited engagement with the child and childhood. In the early 1980s, Chodorow and Contratto (1982, 70) argued that "feminists have been trying to build a theory of mothering without examining or noticing that a theory of mothering requires a theory of childhood and child development as well." They suggested that "feminist theories of the family would both enrich and be enriched by more careful attention to childhood itself and to other age-defined stages of life" (ibid., 10). Barrie Thorne (1982, 9-10) pointed out that much feminist theory assumed "the structuring of generation" and that "feminists have emphasised gender much more than generation." She (ibid., 2) suggested that feminist analysis of the family, attending to the structures of sex, gender, and age, would make "a significant contribution to rethinking the 'naturalness' of the family and to recuperating it for social and historical analysis." Nearly a decade afterward, Sarah James (1990, 281) repeated that "it is difficult to talk about the liberation of women without also including children." However, twenty years later, despite these interventions, Erica Burman and Jackie Stacey (2010, 227) had to reiterate that "the subject of childhood has not featured as prominently within feminist debates as it might have done." Thus, eco/feminism *and* feminism would both benefit from a concomitant attention to age and generation, to the rethinking of *childhood* as well as motherhood.[1] The *naturalization* of childhood, as well as the conceptualization of the boundaries between childhood and adulthood, is key here.

Even if feminists have not always addressed childhood as thoroughly as they might have, as the work cited above demonstrates, there have been a range of important interventions. Smart, Neale, and Wade's *The Changing Experience of Childhood* (2001), which identified four common models of childhood, all of which rely on ideas about a child's innate "nature," is useful for addressing the naturalization of childhood. The models they identify are not mutually exclusive and often overlap and are combined in both popular and "expert" depictions of childhood. In the first model the child is inherently bad, a little devil in need of harsh treatment. In the second model, the child is wild and savage rather than bad, in need of the civilizing influences of adults or of acquiring civility through socialization and the journey to adulthood. Conversely, in the third model the child is angelic, inherently good and innocent (drawing on Rousseau). And fourth, the "embryonic" model draws on developmental psychology, and although providing an account of change, rather than fixity, of a child's nature, is still an essentialist model. Children are neither inherently good nor evil, but rather emergent, unfixed and fragile, although with the capacity to become bad, wild, or good. As Smart, Neale, and Wade (ibid., 3-4) noted, "In this evolutionary model, children start off as little more than simple biological organisms or blank slates. They then progress, both physically and intellectually, through a series of sequentially linked phases or stages through their childhoods ... until they eventually achieve the fully social state of adulthood." The evolutionary model brought together the biological with the social, and explained how the child could be civilized to avoid becoming bad, being always inherently good, or just wild. In this model, children are valued more for what they will become than for what they are; they are not thought of as persons, and personhood is equated with adulthood.

In challenging the view of children as pre-social nature, Thorne argued that they should be seen as agents, along with women. She (1987, 105) pointed out that children are "human actors negotiating with and sometimes resisting institutional structures – schools, families, wage labour," and I would add, environmental destruction. Thorne (ibid.) radically concluded, "If our theorising began with selves defined through relationships with others, retaining full awareness of social hierarchies, and if that approach developed into a full critique of existing institutions, we might thoroughly overturn traditional knowledge." She suggested that including the child in the domain of the social, and in social theory, could be accomplished by refiguring the very concept of the social through a notion of differential relationality. Nonetheless, and despite Thorne's early interventions, naturalistic models remain powerful. Claudia Castañeda (2001, 47, emphasis in original) identified their persistence in feminist theories of the subject and maintained that such theories "return to the child *as* a space that is, in a sense, free, or at least more free, from this subjection." For Castañeda (ibid., 47), "In feminist theorizations of the subject via psychoanalysis, the (girl-)child is the ground of subject-making, that which comes before 'us' as subjects; the child is our 'nature' before culture, the body that we occupy in order to imagine our making." There remains much eco/feminist work to be done to counter the naturalization of childhood.

From the Nature of Children to Children and Childhoods in Nature

It is the association with (non-human) nature that often compounds the naturalization of childhood, as in the notion that rural spaces are the ideal location of an idyllic childhood. Childhoods, particularly rural childhoods, have commonly been celebrated as a time of innocence, of closeness to nature, of communion with animals and other creatures, of wild abandon with hoards of other children, of health, fresh air, and physical play. Furthermore, such representations of childhood implicitly portray adulthood as expulsion from Edenic childhood and initiation into the world of culture. Adulthood is then about the loss of innocence, especially sexual innocence, and childhood can be fetishized as a time and place of nostalgic longings.[2] Thus, the boundaries between childhood and adulthood often map onto the boundaries between nature and the social/cultural.

Although relationships between eco/feminism and feminism are the main focus of this book, here I want to discuss eco/feminism's relationship with (versions of) deep ecology. This relationship has certainly been contentious at times, with debates about population and reproduction a central focus. Misanthropic strands of deep ecology have argued for population control, insisting that humans are a burden on the earth's resources and that future populations should be limited. Earth First! co-founder Dave Foremen notoriously claimed that AIDS was a "natural" response to overpopulation and famine in Africa (Seager 1993). In contrast, other strands of deep ecology have reified the figure of the child and especially childhood relationships with nature. This is particularly interesting because eco/feminists and deep ecologists have had many tense encounters over misogyny, misanthropy, and population, but eco/feminists have paid little attention to the naturalized and romanticized male child who figures strongly in much deep ecology literature.

While academic eco/feminist literature has paid little attention to children as activists, significant evidence from Clayoquot and elsewhere demonstrates children's own environmental interest and activism.[3] For instance, many who became involved in the campaign did so precisely because their children voiced concerns about environmental degradation. Kim Back's son worried about the future of both the planet and his own children. Jane Fawkes mentioned that Mary Sommers, a long-time peace campaigner, was prompted to return to activism by her grandchildren's questions, despite a terminal illness. Betty Krawczyk's daughter, Marian, then in her late teens, had intended to be arrested but Betty took her place, suspecting that being arrested and charged would have a negative impact on Marian's future, including her employment prospects. Children spurred their parents to activism, were arrested with their parents or with the support of their parents, and were sometimes more willing to risk arrest and trial. Furthermore, support from children, including adult children, was often important to arrestees; a number were disappointed when it was not forthcoming or when children disproved of their activism.

Jean McLaren's account of the camp *Spirits Rising* (1994, 25-27) included a chapter in which she reflected on children at the camp and her engagements with them. Children's presence at the camp manifested in many ways. Adults formed a childcare group to enable small children to spend time with each other and to give parents and carers a break from childcare. Young people voiced their concerns about deforestation as well as the possible opposition of parents to their decision to be at the camp and to be arrested, in the conflict resolution sessions that she ran. There were also signs of more overt organizing. A fourteen-year-old boy set up a children's circle – an implicit acknowledgment that children did not always find it easy to speak up in the camp circles and that their views were not necessarily sought or considered by adults. One particularly powerful intervention was when children planned their own day of arrests at the blockades. When the RCMP initially refused to carry out the arrests, the children and young people involved insisted that they be arrested. When they arrived at Ucluelet jail, their parents were told that social services might take custody of them. Ultimately, this threat was not carried out, but parents and children spent hours worrying about what might happen. Many people perceived the RCMP response as an indication of the power of children to resist authority, and the difficulties for the police in coping with such a challenge. During her interview with me, Irene Abbey mentioned her concern that, as one of the oldest arrestees, she received more attention than young people did, despite their possible sacrifices, not least the consequences of a criminal record for future employment.

When children are discussed in eco/feminist literature, they tend to be presented as victims of pollution and environmental degradation, providing an impetus to activism. Karen Warren (1997b, 11-12, 19n47) cited a variety of statistics to show the high numbers of children at risk from lead poisoning, asbestos, and cancers. Stressing that the situation worsens for children who live in poverty, for African American, Hispanic, and Asian American children, and for Navajo children and teenagers, she also referred to the effects of deforestation, desertification, and soil erosion for women and children in the Third World. Ruthanne Kurth-Schai (1997, 193) noted the tensions in reflections on children and childhood in the environmental imagination, observing that "though the image of children as hope for the future is shared across cultures and continents, the experience of childhood as a time of innocence, security, self-worth, and contribution to family and community is a distant fantasy for most children." While she discussed children's vulnerability to pollution, Kurth-Schai (ibid., 207) also pointed to their potential to generate "utopian social and political imagery" and to engage in environmental activism; she thus highlighted the importance of creating space for children's political voice. Critical of protectionist or liberationist approaches to children, which see them primarily as vulnerable victims in need of protection, or as threats, or as learners, Kurth-Schai (ibid., 204) advocated an ethic of pluralism, which would recognize that "there is no singular experience of childhood," and she was critical of universal stage theories, which are popular in some deep ecology texts. Yet Kurth-Schai's argument, like that of Warren, lacked attention to the gendering of children's lives.

Unlike eco/feminism, deep ecology and other environmentalisms have not neglected questions of children and nature. The child in nature is a common trope in much deep ecological writing, where childhood appears as the origin of (adult) ecological concern and/or as a site of an original and unspoiled relationship with nature, which has become tainted by the problems of adulthood – children have exemplary relationships with nature that adulthood disrupts. Thus, adults who attempt to get in touch with nature are getting back to innocent childhood natures. There is also an extensive deep ecology literature on children's relationships with animals, which I do not explore fully here, although its themes are similar (see, for example, Louv 2009). The key questions for all of this literature are whether such childhood natures are romanticized, to what extent gender is considered, and how the boundaries that separate nature from culture, child from adult, female from male, private from public, and domestic from wild are configured.

Much of the literature on relationships between children and nature examines the significance for adults of childhood memories of nature, pointing to the formative role of outdoor play in developing an ecological sense of self. Edith Cobb's work, which links adult creativity with childhood imagination and play, particularly play in nature, is a key reference here. Cobb examined three hundred autobiographies of creative thinkers and claimed in *The Ecology of Imagination in Childhood* (1977) that environmental experiences in childhood made lasting contributions to adult creativity. Louise Chawla (1994) subsequently pointed out that Cobb drew universalist conclusions from her work. Most of the autobiographies that she had studied were written by nineteenth- and twentieth-century British authors whose work was infused with Wordsworthian romanticism in which children were believed to experience a special sense of wholeness with nature that left a lasting adult impression. Perhaps because of this romanticism, Cobb's book has appealed to many deep ecologists who remain convinced of the significance of childhood relationships with nature. For instance, environmental educator Mitchell Thomashow drew on Cobb in developing his versions of "ecological identity" and "ecological self." Invoking the universal stage theories of child development, Thomashow (1995, 10, emphasis in original) claimed that "from the perspective of human development, the period of middle childhood (the ages of 9 to 12 years) is a time of *place-making* in which children expand their sense of self." He characterized this stage as one of exploration and independence, when children establish their connections to the earth, which are crucial for developing their personal identity. His citation of educator David Sobel is instructive: "Children leave the security of home behind and set out, like Alice in Wonderland and Columbus and Robinson Crusoe to discover the new world. In the process, children create new homes, homes away from home. These homes become the new safe place, a small world that they create from the raw materials of the natural world and their flexible imaginations" (quoted in ibid., 11). Here Columbus and Robinson Crusoe personify the "colonising tendencies" of the deep ecological "expanded Self," much critiqued by eco/feminists (Kheel 1991, 64).

One fundamental aspect of much deep ecology writing is the romanticization of home, rootedness, and childhood (Cuthbertson, Heine, and Whitson 1997). Thomashow (1995, 8), for example, wrote, "I've come to know that most people, regardless of their culture, are comfortable talking about their ancestry or their roots. They also like to talk about the places where they live, especially when they are very far from home." He attempted to attend to diversity, referring to the various original homes of his students – an Israeli woman whose father escaped the Holocaust to come to Tel Aviv, a Tanzanian man whose recent ancestors were a "hunting-gathering tribe of proud warriors," and a Nepali man whose childhood forests had been destroyed. Nonetheless, Thomashow (ibid., 9, emphasis in original) proceeded to erase the specificity of his students' backgrounds: "Despite the great variety of international and cultural experience, there is a striking thematic pattern; whether the person is from an Asian tropical rainforest, an African savannah, a Latin American city, a European valley, or a North American farm, they tell a similar story. They have fond memories of a *special childhood place*." He (ibid., 12) did acknowledge that "not everyone can return to those place-making years. For some these memories can be indistinct or painful. Perhaps they experienced a great deal of turmoil because their family was breaking apart. Or perhaps they lived places they would rather not remember." Yet this argument seems contradictory – on the one hand, people enjoy talking about childhood places, but on the other hand, not everyone wishes to return to them. At the same time, Thomashow (ibid., 9) juxtaposed his romantic construction of childhood as a time of special places with "a distressing contemporary picture" of (adult) experiences of "disturbed places." Thus, distressing childhoods are expunged from his account, and childhood and the past are represented as a site of nostalgia, with adulthood marked by expulsion from paradise.

These accounts pay little attention to gender. Although drawing on Cobb's work, Louise Chawla did scrutinize gender in understandings of experiences and memories of home and nature. In examining the role played by memories of childhood nature in the creativity of professional authors, Chawla (1994, 494) identified how "childhood memories of home [and I would add, relatedly, nature] express the internal family relations that the house as a building contained and the external social context within which it was contained ... In each case family relations pervade vivid images of the childhood house. When parents have suffered from political upheavals and economic or racial vulnerability, the children have inherited their struggle to secure a sense of home." Furthermore, Chawla (2002, 481) explored how the "quality of family relations and external social, political and economic conditions in childhood homes influence adult meanings of home." Her work, which provides a critique of deep ecology and begins to suggest fruitful departure points for an eco/feminist approach to childhood natures, informs my approach to the following stories. I turn first to Betty Krawczyk's account of her childhood.

Remembering Childhood Natures

Nature as Louisiana Swamp: Betty Krawczyk

Betty Krawczyk was one of three grandmothers who were detained by the RCMP on the first day of arrests at the Clayoquot camp. Alongside Judith Robinson and Inessa Ormond, she refused to sign the statement that would have required her to agree not to return to the blockade. All three were held in jail over the summer, with Betty remaining the longest, for four months. When I first met Betty, three years after this, she was working as a volunteer in the Friends of Clayoquot Sound office in Tofino. Since her release from jail, she had written *Clayoquot: The Sound of My Heart* (1996), which centred on her involvement in the campaign. Most of the stories that she initially told me were in her book, oft-told narratives that she had reflected on and presented in a way that revealed as much as she wished. She seemed to take pleasure in recounting them, playing on the image of "a perfectly respectable if somewhat eccentric grandmother" who had had four husbands and eight children along the way, and now a criminal record (ibid., 5). Her childhood in rural Louisiana was central to many of the stories she relayed about herself. She told me about growing up and about playing in nature, which was a Louisiana swamp. Her account evoked the complex entanglements of her childhood landscape – social, political, and physical, as well as the relationships between them:

> The area that I was raised in was outside Baton Rouge, Louisiana, and it was on the far side of East Baton Rouge Parish, and it was close to a swampy area, and we lived in the country, we were country people, and we had, my brother and sister and I, had the run of wonderful swampy woodlands. We roamed around in the woods and were conversant with the woodland creatures, and in Louisiana there's, at that time anyway, there were a lot of animals, there were, the swamp animals, which is, frogs and alligators and crawfish, and then of course there were the racoons and the possums and the deer, and all manners of creatures, that, sometimes we ate them – I never ate them – and sometimes we were just aware that they were there and part of our environment. But my mother and father both loved the country life and we, I spent a lot of our time outdoors, so that my earliest memories are of being outdoors in Louisiana and of the natural surroundings of the woods and the creatures that lived in the woods and the bayous, and the water things. There were a lot of bad things too – well they were bad to us – we constantly had to watch for water moccasins and rattlesnakes, there were also poisonous spiders, black widows. Not all the creatures were benign, but they were all part of our landscape, and my brother and I – my sister wasn't a terribly outdoorsy person – she was quite a bit younger and she was frail, sort of allergic reaction to the sun, so she wasn't out very much. But my brother and I spent a lot of time outdoors, exploring the outdoors, and my happiest memories are of the outdoors, and my childhood in this sense was very happy, because, my brother being the oldest and a boy, he got a lot of attention, and my sister, because she was the baby and frail and she got the rest of it, and I sort of didn't get very much, but that worked out really good because you didn't want your father to pay any attention to you anyway, because when he did you were in trouble, and so I had a lot of room just to roam and learn about the natural world, in the sense of incorporating it into my consciousness, and my childhood was very happy in that sense. My father was a very patriarchal man. He was a Southerner to the bone, and when I was about thirteen he got religion and went to school and became a minister, but prior to that he was, he just was raised with this in his bones, that women were inferior, and had, that their primary purpose was as breeders and as wives and mothers. And it's so peculiar because he loved my mother devotedly, and

> he respected my mother, yet this respect for her didn't translate into the female gender as a whole, and he didn't have much hope for my sister and I ever amounting to anything, and why, whereas my brother did get, some sort of planning was put into his future. But the part of the South that I came from, is, was very backward, very ignorant, was very racist, sexist, and very religious, paradoxically, very religious ... And on the one hand, it was a horrible place in terms of the social life – I can remember when I was in elementary school, well just going into junior high, and my father was one of the ones trying to get the subject of evolution out of the school system ... Yes, I mean, this is the kind of background I had. And women married very very young, I married at sixteen, and that was par for the course. I think that only four of the girls I started high school with actually finished, mostly they all married. If you weren't semi-engaged by the time you were eighteen, your parents were getting kinda worried about you. The courting aspect of it was very rigid, we had absolutely no freedom after you became an adolescent ... Yes, these courting rules were actually very rigid, because of the class ... Southern culture had, there were class divisions within the culture, and even country people from city people, and in the country, if you were a sort of a genteel white class as opposed to a poor white trash class, the one thing that rocked the division between these two classes, one of the main things that marked the division between these two classes was the way that the females were allowed to behave, or the amount of freedom that females were allowed to have, and as we were not of the poor white trash class, that meant that my sister and I had absolutely no freedom once we hit puberty, you didn't, having a boyfriend was totally out of the question. You didn't, you weren't allowed to be around boys at all to speak of, other than at school, and any encounter with a boy had to be mediated through your father, and ... you weren't allowed to associate with black people any more, when you were a child you could play with black kids, you could play together, but once you got to be a big girl.

Betty's depiction of her girlhood as one of endless outdoor play appears to echo popular ideas about idyllic rural childhoods. As Owain Jones (1997, 162) reflected, "Country childhoods are seen powerfully in terms of a synthesis of innocence, wildness, play, adventure, the companionship of other children, contact with nature, agricultural spaces and practices, healthiness, spatial freedom, and freedom from adult surveillance." Many of these ideas featured in Betty's account, which contrasted starkly with her sister's fragility and confinement indoors. Although her "childhood in this sense was very happy," at the same time she also recognized that Louisiana "was a horrible place in terms of the social life," a comment that distinguished between the social and the natural. Her adult self identified the effects of the political and social world on her childhood, which she articulated as prevailing racism, (hetero)sexism, and class politics, all compounded with and through religious fundamentalism.

Betty's account presented a constant tension between the natural and the social worlds, almost as if she had two separate childhoods, one in the swampy woodlands, and the other in the social world. The social world is made meaningful and relevant through the influence of her father, who regulated her behaviour and her expectations of life on the basis of her gender and age. His presence dictated the contours of home, where it was best to be ignored or invisible, and outdoors, which became a place of freedom from his constraints.

In this way, Betty provided a vivid depiction of the lived effects of dualistic discourses of "nature" and "culture" on her life as a young girl. Nature was outdoors, a place of liberty in contrast to the oppressiveness of both her home and social world. Betty's account throws into relief the inadequacy of binary oppositions, demonstrating that the private and the public were much more entangled for her. Even as a child, the private, as in the home, was not a haven from the travails of the public world; indeed, it was the site through which the rules of the social world were enacted. Access to the world outside her home was circumscribed, restricted by age and gender, and at puberty outdoors was no longer available to her in the same way. Outdoors begins to suggest a public world, but in fact the nature of Betty's childhood is distinguished from both home and the social world; nature, and access to it, is defined through a specific intersection of age and gender. If nature as the rural and as the countryside offers children a space of freedom and wildness, we are forced to ask "freedom from what?" From adulthood? From their parents, teachers, and their own impending adulthood, which will define the end of childhood and the end of innocence?

David Sibley (1995, 123) argued that boundary questions are an important focus for research on childhood because "boundary experiences affect the quality of interpersonal relations and the quality of the relationship between the child and the material environment." Although he cited research that addressed concerns about childhood spaces and boundaries, he stated that "nowhere are children situated at the same time in the context of the family, domestic space and the larger spaces of the locality and the city." Echoing many feminist insights, Sibley (ibid., 129) noted that "the home is one place where children are subject to controls by parents over the use of space and time and where the child attempts to carve out its own spaces and set its own times" and that

> much of the literature on the home, particularly in environmental psychology, fails to convey the frustrations and anxieties that may be associated with home life. The dominant message of environmental psychology is that the private domain of the home is a benign, controllable, personal space standing in contrast to the exterior, public domain which is uncontrollable, uncertain and riven with conflict. The house is haven; the dwelling place in western culture is a "locus of sentiment."

However, as feminists have pointed out, the home is invested with power relations and is not necessarily benign for children and women. The constitution of the boundaries that define home, nature, and the public world also mark the boundaries of a childhood, which is as much gendered as determined by age.

In discussing the rural, nature, and the gender of childhood, Owain Jones (1999) argued that the "natural" gender of childhood is male. He contended that romantic constructions of, and connections between, childhood, nature, and the countryside leave little space for female identities; girls who play in nature and in the countryside are popularly represented as tomboys. Although he observed that his interviewees often referred positively to the tomboy, as denoting resistance to other constructions of gendered identity for girls, he also cited prevailing ideas about the emergence of female sexuality as precipitating the end of childhood, marking a departure from the natural state of childhood/innocence. As Judith Halberstam observed in *Female Masculinity* (1998, 6), "Tomboyism is tolerated as long as the child remains prepubescent; as soon as the child is older, tomboyism is punished."

In telling her story and trying to reconcile her contradictory and somehow separate experiences of childhood, Betty invoked many different ideas of nature. The nature of her childhood was very specific – the swampy woodlands in rural Louisiana. She highlighted its particularity, in part through naming the various common creatures that lived there. This was not just any rural area, any countryside, but a specific place that had much meaning for her. In mentioning the animals that share her landscape, she corrected herself because she now lived on Vancouver Island, which was populated by different creatures. It was also a specific time – she shifted to the past tense, "at that time anyway," acknowledging that many of the creatures were no longer so common, that nature changes over time, and especially hinting at the impact of humans.

She sometimes naturalized the countryside and the rural (and the people who lived there). Aspects of "Southern life" were presented as inevitable, unchangeable, as with some characteristics of the people in the area. Her father, a "Southerner to the bone," believed "that women were inferior" and "that their primary purpose was as breeders and as wives and mothers." She said, "We were country people." She distinguished between good and bad natures; water moccasins, rattlesnakes, and black widows are bad creatures, but she modified this anthropocentric view by pointing out her situatedness, that these animals "were bad to us."

She does not recount her childhood through a nostalgic lens as a time when life was simpler and easier; nor does she represent it as a point of origin from which the rest of her life unfolded teleologically. Rather, she constantly returned to it in an effort to understand her experiences as a child and to tease out its continuing significance, but there was no straightforward correlation between her life then and now. Her current understanding of her childhood experiences in terms of class, race, and gender represented an attempt to understand these apparently separate worlds.

Although she left the country behind and moved to the city as soon as she could, it still haunted her, as is clear from her reflections in *Clayoquot: The Sound of My Heart* (1996, 12):

> I don't know what it is with this wilderness gene. If I had known from the beginning what kind of life makes me happy I could have saved myself a lot of misery. I was raised in the country, in East Baton Rouge Parish, in the state of Louisiana. I loved the country, but the aim seemed to be, on everybody's part, both black and white, to get out of it as fast as possible. It's hard to hang on to something that is so down-graded by everybody else. In fact, my brother and I couldn't beat the country dust off our shoes fast enough. Downtown. Just show us the way. The funny thing is, both my brother and I have since spent a lot of years trying to get back to the heart of nature.

> But one can't really go home again ... And while the Clayoquot Sound evokes some childhood memories for me – the heavy rainfall is sweetly familiar, as is the fresh, sea borne air – still most is strange. It's like meeting a new man, the strongest attraction is centred on the one with the most challenging blend of the strange and the familiar. Only my fascination with the Sound hasn't worn off yet. At least not yet.

For Betty, there was no return to "home," to the nature of childhood. There was rather a constant tension between rural and urban living. She knew that the countryside was not necessarily idyllic, that being a girl and becoming a woman was difficult there. Her sense of connection with nature was perhaps most literally conveyed in her reference to having a "wilderness gene." Although this might read like a turn to essentialism through biological determinism, such an interpretation is too simplistic. It is not clear whether she has inherited this queer kinship with nature from her parents, or whether it is a mutant adaptation to growing up in the social world of Louisiana, a result of campaigning against logging, or just a result of too much time spent playing in muck. Genes carry DNA, the ability to replicate life, which also suggests that she may "reproduce" wilderness, passing it on to future generations, perhaps through her campaigning to save Clayoquot Sound, as an activist role model. The future of the earth is linked with that of her grandson – it is not only a question of whether there will be a planet left for him to live on, but also whether he will continue her work in some way.

In recognizing that she cannot go home, Betty must make home in Clayoquot or wherever she lives. Perhaps Clayoquot Sound will never be home, but it is "the sound of her heart." It is strange but familiar; the use of "familiar" is suggestive of "in the family," some unknown and as yet unmet relation to Louisiana, a further branching in her genealogy. Living in Clayoquot requires work, home-work – to become familiar, to familiarize herself with it and its inhabitants, and to engage in the ecological and political work required to sustain it, which for Betty included months in jail. She mentioned two motivations for her arrest. As noted above, she participated in the blockade so that her daughter Marian would not do so. The second motive was her realization that the replanted trees around her cabin were not growing back, as promised by the person from whom she had bought the land, because clear-cutting had damaged the property so extensively that young trees couldn't thrive. So, though Clayoquot is a compelling landscape for Betty, so compelling that she has been arrested and jailed many times, she understands that it will never have the same meanings for her as Louisiana.

Betty's account demonstrated that the boundaries of home, nature, and the public world are important in defining gender and the place of women and girls. The themes that she discussed were also mentioned by other women, who spoke of nature as a childhood refuge, as a source of solace and a place of freedom. Most conceived of nature as "outside" – outside the walls and boundaries of their houses, and also outside the social world, separate from an adult realm of culture. Many spoke about playing in nature or described specific places that they had visited as children, places with meanings that continued to resonate. Interacting with nature in this way did not challenge understandings of (pre-pubescent) femininity, and few talked about relationships with nature that transgressed the acceptable. Many learned about nature through their mothers and also their fathers or grandparents. Louise Chawla cites the importance of childhood family relations in understanding meanings of home – and I add nature and the social world (though Chawla does address these subjects to some extent in her focus on the social and material features of home). For Betty, family relations, particularly her father's dominance, meant that she spent as much time as possible outdoors, but once she reached puberty her access to it was increasingly circumscribed. Ultimately, marriage was the only route out of the family home.

Many women recount childhood natures as a space apart, as separate from home or the social or public world where the rules of patriarchy are learned and enforced. Nature is a place of liberty, where such restrictions have no meaning. For Betty, puberty reduced her access to nature, but access can also rely on whether a girl is alone or with others, such as family members or boys. Boys are often permitted to enter places that are off limits to their sisters. Distinctions can be made between spaces or natures, particularly between urban streets and domestic gardens or rural natures, but they highlight rather than undermine the argument that girls and women are regulated, disciplined, and subjected through defining their access to particular spaces.

Nature as Garden Hedge: Miriam Leigh

Miriam Leigh discussed her childhood in Virginia during my interview with her:

I had two older brothers and often heard, "No, you cannot do that, because you are a girl," and I couldn't understand why and it would make me very angry, because I knew I could do it, there wasn't any physical reason I couldn't do it, and to just be told I was a girl was very enraging. I didn't get to – just, this is a silly aside – I guess one of the big sore points was trick-or-treating. Young girls, girls were not allowed, in Virginia it was not proper at that time for girls to go trick-or-treating in Virginia at that time, so I would stand and watch my brothers on Halloween get all dressed up and go out and about, and have a wonderful time, and finally I think when I was about twelve, this woman who was in the neighbourhood and who was very different in her viewpoint on life, she was divorced and had to work for a living, took me trick-or-treating with her daughters, and it was just wonderful [laugh]. So, that was the kind of cage that we were in as girls, and often I remember fleeing into nature. We lived in a small town till I was seven, and nature for me when I was really little was this huge hedge that was in front of the house that was made of trees, and so I would go inside the hedge, and that was quite magical in there and that was my refuge and I spent a lot of time outside, just either making up imaginary games or playing, but outside was freedom and I still feel that in my life. If I have to stay inside very long, I start to go quite batty. And then we moved to a farm, or out in the country, when I was about seven or eight and there was, in Virginia they don't call it a forest, they call it a woods. We have a lot of deciduous trees as well as some pine, though it was mostly deciduous, and the place where we lived, it had two old pines on it and I spent a great deal of my growing up years around these pines, creating all kinds of worlds and games and refuges. So I guess I have always kind of seen nature as a place of solace and nurturance and protection actually. And my mother was very, very into the woods and would take us on nature walks every Sunday, and taught us the names of all the trees, and the shrubs, and the wildflowers. And was very into conservancy. Both my mum and dad were into stopping kind of the thrust of commercial civilization eating up all the farmland. My dad had – he put a lot of the land into what at that point in time was called conservation – the idea was for, the government pays the farmer to let the land lie fallow, so that the wild animals would have somewhere to live, some sanctuary, the birds especially, and my mum was really, really into saving trees. She just thought that it was terrible to cut down all the trees.

Miriam spoke of her sense of injustice at not being allowed to do the things that her brothers were allowed to do, mentioning one example, of trick-or-treating, which would involve walking urban streets at night and in the dark, a possibility circumscribed by her parents. It was only with a woman who had already transgressed the boundaries of respectable womanhood by being divorced that she was able to go trick-or-treating. Through such processes of regulation and discipline girl children are constituted as distinct from boy children, but there are acceptable ways for girls to go outdoors. One is through walks, particularly nature walks, with a parent. During these, Miriam's mother taught her the names of wildflowers, trees, and shrubs, providing one method of connecting with nature. Miriam had her own way of making sense of nature – the magical hedge in front of her house was a place where she could play and make up imaginary worlds. This nature she experienced as freedom.

Miriam's account points to how natures can have different meanings, and how experiences of nature differ according to contexts, and whether alone or with others. Her reflections also point to the role of parents and families, and their understandings of particular natures as safe or dangerous, in determining childhood experiences of nature. Some spaces are deemed appropriate and accessible for girls, whereas others are not; Miriam's brothers can venture into the urban streets, but she cannot, though she was allowed to play endlessly in the trees in front of the house (perhaps within sight of her parents).

A Farming Childhood: Judith Thompson

During my interview with her, Judith Thompson recalled happy childhood times spent in nature, but she also introduced the issue of landownership in determining where adult women could live and how they could support themselves:

> My experience with nature was through my grandparents, who have a large amount of land in southern Ontario, where I spent all my free time. My mother and my sister and brother and I and my dad spent a lot of time in the country and on the water and southern Ontario's Great Lakes. And my grandfather was one of the first conservation officers in the small town of southern Ontario where he grew up, and so there is a tradition that I learned from him and I learned to fish from him and to go out in the woods

and be quiet and listen and respect nature. He had a deep respect for nature and he had a deep respect for Native people and their wisdom. I was a fairly lonely child, so I spent a lot of time by myself in nature and just observing, just that, you know, being able to sit and watch and listen. I never was keen on the interventions in nature, and I was not very keen to kill anything or even to fish. My joy of going fishing with my grandfather was the stories and just taking a look and listen and be quiet in nature. So that was a huge part of who I am – and is a huge part of who I am ... I felt much more comfortable in nature and much more myself than I did in those social situations where I never did understand the rules and when I did grow to understand a lot of the rules – the tyranny of some of the rules – I fought against it strongly. One of them, of course, was patriarchy and what happened to the family farm was that even though my mother had been very involved in the farming and driving tractors and doing pretty much of the same work as men, when it came to who was going to get the family farm – a woman could not have it. And so we lost the family farm. Basically because my mother was the only one who would have been able to have it. So that was a heartbreak ... And when I first came to BC I was twenty, twenty-one, I think, and I immediately went to Tofino and lived on the beach ... in the sixties when everybody moved out to the land and those were wonderful times. I was really quite confused in my early twenties as to who and where I was going, because I didn't feel like I ... My graduating class: most of those women were engaged and married and already starting families by the time we were just graduating. I knew that was something I didn't want to do. So I travelled ... and then, I guess, moved to the west coast and lived in Tofino for quite a number of years, just being totally in love with the ocean and the richness of that rainforest. So I was twenty-one years old and I wandered and camped in Clayoquot Sound. So that's now thirty years ago, when there was no logging in Clayoquot Sound, when the pristine beauty was just such a spiritual experience, I can't even explain the kind of peace that brought to me and I was just confused in my young womanhood. And I met some incredible women on the coast who were living basically in isolation, managing their own food stores, extremely independent. I learned a lot about gardening, learned a lot about harvesting from nature. Well, it was just an incredible experience. I lived on the beach for about two years, I guess. But Clayoquot was very dear to me. That was my university. That's where I learned everything about the world and myself – what a challenge is and what freedom is. Well, I wanted to have my children be able to do that. I'm not so terribly concerned

> that they're getting four years of university. I'm really concerned that they have the choice to go into the ... forest and just learn about nature and learn how things interact and to be a part of that. So that's why I went to Clayoquot basically. I went with my son ... He was thirteen.

Like Betty and Miriam, Judith evoked a specific childhood nature – around her grandparents' home in southern Ontario and the Great Lakes. Like Miriam, she experienced nature through her companionship with family members, specifically her grandfather. In contrast to Betty's robust playing in nature, her connection was grounded in quiet observation. Nature was a separate place, where she could feel "much more herself" – it was distinct from the social world and its incomprehensible rules, which left her confused or rebellious and which she later identified as patriarchal. But, because they also determined landownership, the rules circumscribed her experience of nature. Her mother could not inherit the family farm, and her own wishes about where and how she wanted to live were not taken into consideration. She came to Clayoquot in her early twenties, again confused by social expectations, this time that she should marry and start a family. Nature was a place of healing for her, and the independent women she met at Clayoquot taught her how to live on, and with, the land. In Clayoquot, confusion gave way to understanding, and she learned "everything about the world and myself." Returning to the area thirty years later, drawn by the need to protect the place where she found "peace," she was shocked by the destructiveness of clear-cut logging. She went to Clayoquot *with* her son rather than *for* him, wanting to share something of herself and her life with him.

Isolated Natures: Sophie Rogers

During her interview, Sophie Rogers skimmed over the early part of her life and talked more about her time in the countryside: "I was born in Calgary and lived in the city until I was eight, and then moved out into the countryside to a kind of farm, not quite a farm, we had a bunch of animals, mostly dogs, cats, and horses and chickens and a huge garden." She did not remember much about the city, but she told vivid stories about her life in the countryside:

> I remember we used to build tree-forts. We lived right by a patch of poplar trees, which I thought at the time were big until I moved to BC, and so, I, we, built tree-forts, and it was like a, sort of, a place to hide when you

> got into a fight with your family, and it gave me a place to hide, and I liked that. We'd go out to this tree-fort, which I did a lot and sit there thinking, and I'm not sure that my thoughts were the same as they are now, but now I think that I often wanted instead to have a person to talk to, instead of going out and hanging by myself, because that didn't give me a whole lot of life answers, y'know, it removed me from the problem, so I think I was conscious of being very removed from other people, and I didn't have very many friends, close, in the neighbourhood. So that was a difference. My friends who lived in [the nearest town], their friends were close by.

For Sophie, nature offered solace and nurturance but not in the sense that it did for Miriam and Judith; it was a place of isolation and loneliness where she hid from her family, a place apart from friends and human contact. Sophie's account of nature was more ambivalent, pointing to the problems of isolation for rural residents, particularly children and those who rely on others for transport, an important issue in Canada, given that rural areas are sparsely populated, with more than 90 percent of Canadians living in urban areas. However, though Sophie missed regular contact with her friends, she also hated going to the mall with them, preferring to spend her time on horseback: "I loved horseback riding and I think a lot of what kept me a sane person was horseback riding."

Sophie grew up in a fundamentalist Christian family, which she described as highly controlled. Her childhood experience of nature, like those of Betty, Miriam, and Judith, was strongly influenced by her family. For example, when the family moved to the countryside, she gained experiences of nature, particularly through vegetable gardening and riding. However, she was abused by a minister from her family's church. When she revealed this, the family reacted poorly: "What pushed me out was that behaviour, and telling my mother, and having her say, 'How dare you.' I got strapped for telling my family about this." This moment of not being heard was a key point in her life: it "just turned me off, just about everything I had ever learned, which included growing vegetables and picking wild fruit." When she was about fourteen, her sister was forced to leave home (she didn't explain why, but perhaps it was because she had a boyfriend). The sister moved to a nearby town, and although Sophie did not describe their relationship as close, watching her sister being "exiled" by the family was also "like a pretty huge push." Shortly afterward, she started to run away, eventually contacted her sister, and then started to run to her sister's place. Her behaviour provoked family conferences about what to do with her, where everyone gathered, including the sister and her boyfriend (eventual husband), as she kept running to their home. At these meetings, she was (again) silenced, but her sister's boyfriend spoke up for her and supported her, saying that she had reasons to be upset and that her opinions deserved to be respected. Yet ultimately, he too betrayed Sophie's trust and sexually abused her.

At age fifteen, she ran away permanently, went to Calgary, dropped out of high school, moved in with some friends, and got into smoking, drugs, and alcohol. She became involved with a biker group and started a relationship with one of its members. Reiterating Betty, Miriam, and Judith on marriage, she had the idea that "you are still supposed to get married," and eventually she got engaged to the biker. However, she "decided to do a life change," so she broke off her engagement and took a job as a telephone operator at a hotel, a move that she described as an unconscious attempt to change her life, although she subsequently became a "workaholic" and an "alcoholic." After two years, she developed severe stomach problems, growing sicker and sicker, until eventually she "did another life change" and moved to Vancouver. She said that "either I was going to stay in Calgary, either I was going to stay there and die, which is what happened to an awful lot of my friends from high school, and I just couldn't figure out how to start something new while I was still in Calgary." When she arrived in Vancouver, she "started life number three, and literally put on a totally different face, and left behind all my pretty clothes in Calgary and became a hippie apparently." Since then, she had made no further major identity changes and said that she was trying to consolidate her life in Vancouver.

Soon after arriving in Vancouver, Sophie began to volunteer with the Western Canada Wilderness Committee, an environmental organization with an emphasis on education about wilderness issues. She dated a committee member who "dumped" her because she "didn't want to get married and have babies." Something about the Wilderness Committee bothered her, but she volunteered for quite a long time before she identified it: "It's incredibly male dominated and it's all about control, it's the same shit that I've felt with all the men and all the religion in my life." Being in Vancouver, in a different place, enabled Sophie to get in touch with and transform some elements of her childhood, such as her experience of nature. However, this reconnection was not an exercise in nostalgia but an attempt to change her relationship with nature. It was no longer about isolation and loneliness –

instead, it was about hiking with friends and working with disabled children at a riding school as well as with environmental organizations. In this sense, her activism involved a transformation of her childhood relationship with nature.

Romanticizing the Nature of Childhood?

For many of the women whom I interviewed, memories of childhood natures were central to their activism. This was not a return to childhood. It was not easy to romanticize and sentimentalize these natures and homes. Childhood was not rendered as a site of nostalgic longing for innocent play in unspoiled nature. For all these women, family mediated experiences of and feelings about nature, and the relationships between home and nature. For many, learning about the boundaries and limits of home and nature often entailed learning about gender and what was expected of them as girls and later as adult women. Their stories powerfully demonstrate that childhood, and particularly childhood for girls, is a site of constraint and conflict, where nature is used to mark the boundaries between girl and boy, child and adult, girl and woman, and is invoked by adults in the gendering of children. Girls learn to be girls through the differing access to nature for girls and boys, and nature was invoked to set limits on their lives, to keep them in place, in the domestic realm of the home. For the interviewees, becoming involved in Clayoquot enabled a process of refiguring and transforming relationships with childhood natures, allowing the resignification of the meanings of gender and family, home and nature, as well as the relationships between these.

8

Unnatural Histories

Mother Nature, Family Trees, and Other Human-Nature Relationships

Coming to Clayoquot Sound: Routes and Roots

The women who came to Clayoquot came for many reasons, entangled in other travels and journeys. In some cases, their routes led to roots – they arrived in Clayoquot and planted themselves down (cf. Gaard 2007). Home was not where they came from: it was what they found, or made, once they arrived, not so much a place as a relationship to be made.

Jan Bate came from Ontario in 1988 and eventually became a Friends of Clayoquot Sound (FOCS) director. During her interview, she recalled her process of becoming involved in the FOCS:

> The campaign, the campaign has been, I guess the campaign has been in the papers since Meares Island, which was '84, '85. I was living in Ontario then. I didn't know enough about the west coast of BC, I thought Meares was part of the Queen Charlottes, Haida Gwaii ... and when I actually showed up here I was like, "What, are we that close to the Charlottes?" I didn't really come here because of that. I'd always wanted to live on the west coast ... moved to Vancouver and took trips up to Long Beach, and driving into Tofino it just felt like home. There is, like, no explaining it, it was just like, very overwhelming feeling, and my friends and I were going to go whale watching and we ended up at the local bake shop and meeting all these local protectors who were on their way up to Sulphur

> Pass, so, I was like, I didn't know anything, old growth, clear-cuts, nothing, and so we jumped in the boat with them and went up there and that was eight years ago, so I was already feeling like the place was home, and after spending the evening with those people and getting up in the morning and going on the road, and I think there was thirty-six of us on the road that day, and one person got arrested, and we might all have got arrested that day, but it just didn't work out like that that day, the police were so slow, and hours went by, and gradually people left, and it didn't happen, but, just, I just had this realization, y'know here's a place that I love, and it's filled with people that are dedicating to protecting it, so maybe, maybe it's, like, my future looks good, to put roots in here, and, I mean anyway, it was just the feeling of the place that brought me here, but there was that extra comfort level of knowing that the protectors were really active, and so I stayed.

Valerie Langer came to Tofino in 1987:

> My parents have been coming here for about fifteen years and I have a friend who lives out here or a couple of friends who live out here, and I was on my way to do a tree-planting contract on North Island. And they said, "Why don't you stop in and visit our friends?" so I said fine and I landed up here. And ended up never leaving – I'm one of those many people in Tofino who came for a week and been here now nine years. You'll find lots of people who came for a day and now have been here for twenty-five years, that's pretty common around here. And I got here and I thought this is the most beautiful place I've ever been, why leave? Why go plant trees in dead clear-cuts? And [I] ended up teaching. I'm a literacy worker and ended up doing home schooling for some kids on one of the islands out here and pretty well immediately got involved in the Clayoquot Sound campaign. My parents' friends were Friends of Clayoquot, and I come from a background – my parents are social activists and a sister who is an environmentalist, and I had worked for an environmental organization in Ontario before coming out and it was for me a pretty natural transition to work with a local group on what at that time was a very – one of those really obvious issues like you don't kill babies, you know, that kinda thing, it was these enormous clear-cuts across sandstreams ... Nine years ago, we were dealing with a Socred [Social Credit Party] government, a very conservative right-wing government, massive logging corporations and what

> they called "sympathetic administration," which meant that they got to do whatever they wanted because they could just justify whatever they wanted to do saying "It's economic hard times," so we were in a recession that time, yeah, I got involved because for one the issues were very obvious and this town in particular – it's almost impossible to separate yourself from the environment as you can in most cities. You know, everywhere you look around in Tofino you see the mountains, you see the forest, you see the ocean, it's a narrow peninsula, so even if you wanted to get away from it, you couldn't; you'd have to drive to Vancouver pretty well, so being in this town – it's a politically active town for its population base and it's got a politically active population – and at the time I came here, it was a little volunteer community environmental organization, anybody that, who had an hour to spare – and I just thought you gotta ... put my roots down somewhere and I started here!

Amy Simpson came to Tofino in 1993:

> Before I had heard anything, I was living out in Ontario. In the summer of 1992, I lived in the Rockies, just inside the BC border, and I took a trip out west to travel about a bit and decided that I wanted to move out to the coast, so I went back to Ontario for four months or six months, and then I moved to Tofino, and I didn't know anyone there. I just moved there because my friend from Ontario had lived there the summer before, and she said that she thought I would like it, and she could set me up with housing and a job. So I didn't, I hadn't heard anything at all. I knew that I had heard sort of bad things about blockades on Meares Island, and I had actually been avoiding the environmental movement because of racism and sexism and stuff like that. And at that point I had, sort of analyzing my situation as a woman had been very important to me, probably I think all of my life, but I think, consciously and deliberately since I was about fourteen I think, and at this point I was twenty when I went out to Tofino. And I had been involved in, the first sort of direct action I had been involved in was organizing around the Gulf War. In January 1991, I was part of a youth group, some people organizing at the university, mostly vigils and a sort of a benefit, my first, that sort of organizing work. So when I moved out to Tofino, I was planning on learning how to surf and planning on having a good time, and I didn't really know where I'd be going after that, and two months after I moved to Clayoquot Sound I became involved

> with the Friends ... What led to that – well, it's kind of an embarrassing story, I think I had been involved with a man for three years when I moved out to Tofino, and I think I was starting my coming out process. Well, I think I had started it a couple of years before, but I knew that I was in some pattern, so I went to Tofino and immediately became involved, and he was in the FOCS ... This is probably not what you were expecting – it's kinda ironic though.

Betty Krawczyk (1996, 145, 155) first came to Tofino more than twenty years before the blockades began:

> Perhaps you should get out some, he [Betty's husband, John] suggested. French classes were starting up and did I still want to take an evening course in conversational French? He would watch the girls a couple of evenings a week if I could get them ready for bed before I left for class. Alright. I knew I should get out, do something. I actually realized at this point that the climate [in Ontario] was not good for me, but I had no idea that this misery was compounded by the fact that I was in mourning ... that I was grieving deeply for the loss of the south [Louisiana], for the familiar forests and streams and wild woodland animals that I knew, for my friends and relatives, an entire way of life, grieving too for the confessional stories that I would now never write, and most of all, for the notion that I had a satisfactory marriage ... It was a week later that I accidentally ran into Wally Krawczyk at the post office ... I went home and told John that I was in love with Wally and wanted a divorce. Yes, I expected him to be upset. Yes, I expected his pride to be hurt. Yes, I expected him to bully and threaten me. I no longer believed that he loved me, but he was used to me, used to me being there for him. I expected him to raise a bit of hell, but I didn't expect him to go out and buy a gun and threaten to shoot Wally. And in the process so upset the college which was getting ready for year-end exams, not to mention my children who stared at me with wide, worried eyes, that we all descended into hell for three whole weeks.
>
> It was at this point that Wally and I decided to run away, taking only the children and a few clothes with us. There was no question in my mind where I wanted to go. In his letters Mike [Betty's son] said the south-west coast of British Columbia was a lot warmer than anywhere in Ontario. So one morning after John had left for work I packed a few things for the baby and the older children, called a taxi, met Wally who had rented a car

in another town and was waiting at an appointed place, and then went to school and collected the girls ... Once on the road to Tofino we turned off at Long Beach for a rest break. When I actually stepped out from the foliage of the dense woods onto that beach I stopped in my tracks, rooted in astonishment.

I was familiar with some well known American east and west coast beaches but I had never seen anything like this before. I tried to take it all in with great gulps of fresh, salty air. Miles and miles of wide, hard-packed white sandy beach. The foamy, surging surf splashing and breaking against the giant boulders that ringed the shore line. Old growth rainforest then crept down from the mountains to the very edge of the sand. And as we stepped forward to explore, still stunned from the heart-stopping beauty, we found an incredible abundance of sea life right on shore – clams and mussels and enormous barnacles just underneath the sand or attached to the rocks. We were the only humans on the beach but life was everywhere and everything was breathtakingly beautiful.

And it was warm. Evening was coming on and the air was still warm. Lagging a little behind the others I knelt down and kissed the sand. Oh, merciful Goddess of the universe, thank you. Thank you for giving Canada at least one warm spot where I can survive and for making it so unbelievably magnificent. Thank you. Thank you.

A key feminist insight is that home is not always a safe place for women. As Caren Kaplan (1994, 194-95) wrote,

> We must leave home, as it were, because our homes are often the sites of sexism, racism, and other damaging social practices. Where we come to locate ourselves in terms of our specific histories and differences must be a place with room for what can be salvaged from the past and what can be made new. What we gain is a reterritorialization; we reinhabit a world of our making.

This insight seems crucial for appreciating how eco/feminism might be remaking home. In her exploration of biographies of women activists, Carolyn Merchant (1981, 39) noted that "juxtaposing the goals of the women's and ecology movements leads to important conclusions regarding the future of life on earth. Unless the home is liberated from its status as 'women's sphere' to that of 'human habitat,' the feminist movement cannot succeed." It is not so much the case that activism is a defence of hearth and home, but that becoming an activist often entails a redefinition of women's sphere, breaching the boundaries of both – a reterritorialization.

Unsettling Clayoquot: Arborescence, Rhizomes, and Genealogy

Catherine Nash (2002) has done much to trace the contradictory shifts and connections between "roots and routes." She has examined the search for alternatives to fixed, essentialist, and exclusive ways of imagining culture and belonging, which highlight attention to "cultural routes, mobility and hybridity," while at the same time noting that discourses of indigeneity, of new migration patterns as well as cultural globalizations, have prompted a growth in family history. Nash (ibid., 32) observes that against talk of mobility, flows, travel, and migration,

> the language of family trees, roots, branches, ancestral lines, and origins in genealogy seems to rely on the conceptual structures of Western epistemology that Giles Deleuze and Felix Guattari call arborescent thinking – the Western tradition of rigid, binary logic, and systematic and hierarchical branches of knowledge rooted in firm foundations. Their rhizomatic alternative emphasises shifting interconnection, alliance, networks, fragmentation, combination, and plurality. Roots are certainly out of favour. For them "There is always something genealogical about a tree. It is not a method for the people. The rhizome is an anti-genealogy."

Turning eco/feminist attention to this opposition of roots and routes, sedentary lives and mobility, suggests the ongoing repetition of nature/culture dualisms, where roots are figured as passive and fixed, unlike rhizomes, which are presented as having an agentic cultural life. Against the assumption that trees are stationary, we might recall the generative capacities of seeds, and that in dispersing their seeds, trees take routes and then root all over the world, whereas the networking of rhizomes appears rather more limited in scope. While rhizomes might unsettle notions of being easily able to trace origins, in the ways that trees and seeds might suggest, reproduction and biology may be a little more complicated than that. We might also note the apparent confusion of trees with two-dimensional family tree diagrams and arborescent thinking. Something is always lost in the process of metaphorization, sometimes quite a lot, which is why Haraway's notion of material-semiotics is so useful, in its attention to the intense matter of words. Although Deleuze and Guattari's (2004) argument against what they called aborescent thinking has been widely adopted and sometimes read straightforwardly as a statement against genealogy, this seems to misconstrue their intentions. Although Haraway objected to being read as a Deleuzian, what they might be understood to share is a sense of humour and playfulness. Perhaps we need not take too seriously these juxtapositions of trees and rhizomes, cyborgs and goddesses. After all, like Haraway, Deleuze and Guattari are also opposed to dualisms. We should be suspicious of any suggestion that they propose to reinstate new ones. Their argument for rhizomes as anti-genealogy might best be read in the vein of Haraway's (1985, 101) stated desire that she would "rather be a cyborg than a goddess" – trees and rhizomes might better be understood as bound together in a spiral dance.

As Nash (2002) pointed out, the process of researching a family tree and sometimes travelling to ancestral homes can complicate one's understandings of ancestry and origins. Whereas exploring the self genealogically can be one dimension of a liberal individual self-fashioning and a reflexive engagement with the meaning of ethnic identification, Nash (ibid., 27) is clear that such is not always the case and that "for both individual and collective identities, genealogical projects can have unsettling results." It is with these unsettling roots and routes in mind, and reimagining the possibilities of family trees, that I revisit the matter of maternalism and Mother Nature.

Mother Nature: Metaphor or More-Than-Human Genealogy?

> My physical mother is gone. My spiritual mother remains. I am a woman rewriting my genealogy.
>
> – TERRY TEMPEST WILLIAMS, ***REFUGE: AN UNNATURAL HISTORY OF FAMILY AND PLACE*** (1991)

> The understanding that place is actually our mother, place is what nourishes us, that without this sense of place we're dead.
>
> – DAPHNE MARLATT, 2003, QUOTED IN SUE KOSSEW, "HISTORY AND PLACE"

> I am sick to death of bonding through kinship and "the family;" and I long for models of solidarity and human unity and difference rooted in friendship, work, partially shared purposes, intractable

> collective pain, inescapable mortality, and persistent hope. It is time to theorize an "unfamiliar" unconscious, a different primal scene, where everything does not stem from the dramas of identity and reproduction. Ties through blood – including blood recast in the coin of genes and information – have been bloody enough already. I believe that there will be no racial or sexual peace, no liveable nature, until we learn to produce humanity through something more and less than kinship.
>
> – DONNA HARAWAY, ***MODEST_WITNESS*** (1997)

Haraway (1992, 299, 296) famously challenged kinship as the dominant model for bonding in the late twentieth century, a kinship based on blood ties and naturalized family relationships, writing that she would like "to displace the language of reproduction with generation" and that "we must find another relationship to nature besides reification and possession." Haraway is on the side of the vampires, creatures who leach blood from bodies in the darkness of night, polluting the purity of bloodlines and family genealogies in the process. Her (2007, 231) concern was about "the neat exclusions and genetically distinct, single-trunk trees" of family lineages. Yet, though for her (ibid., 214) "a vampire undoes the family tree and its genealogical method" even as it adulterates bloodlines, Haraway's vampire still maintains the importance of blood in determining relationships. Here I insist that vampires do not constitute the only challenge to kinship and its naturalizing of family relations – Mother Nature is also engaged in this work.

I have reviewed accounts of eco/feminism as mothering nature, as an extension of women's roles from mothering children to mothering the planet. I have also reviewed, especially in Chapter 6, accounts of saving the planet as maternal work – as the naturalization and feminization of environmental labour. Here I switch to a different gendering of nature, the attribution of "mother" to the earth that occurs in pairings such as Mother Nature and Mother Earth. The gendering of nature, especially nature as mother, has been subject to intense criticism within eco/feminism (see Roach 1991; Seager 1993; Vance 1993; and Sandilands 1999). Heather Eaton (1997, 113-14) is critical of "metaphors of the earth as mother, sister, lover or home," which appear to offer reassurance:

> The fact that ecofeminists propose such metaphors indicates a need to tame and domesticate, perhaps even to subdue, the natural world. To use metaphors which imply a soft, kind, receptive healing, warm and loving entity is partial at best. At worst it fosters a utopian, romantic, and imaginary natural world, and hence mars the liberatory potential of any spirituality. To domesticate nature, as these terms insinuate, is to ignore the fact that the natural world not only is untameable but often unknowable and incomprehensibly sophisticated.

I find this interpretation of such metaphors curious, and I offer a passage from Linda Vance (1993, 132), which is quite counter to that of Eaton, to begin to point to alternative possibilities:

> After all, anyone who spends much time in the natural world knows full well that nature is not June Cleaver. I wrote the first version of this chapter while hiking in the forests and mountains of Vermont, where – in June – hail, thunder, rain, and lightning assailed me; where I slipped and slid over moss-covered boulders and slime-covered roots; where I toppled into a crevasse on a mountainside when loose gravel gave away under my boot; and where, at last, I was driven from a rocky summit by seventy-miles-per-hour winds, sleet and snow, and cloud too thick to see through. This nature – my nature – was a wild and rowdy woman, a bad and unruly broad with no concern for her children, and no use to anyone but herself. This is the nature you will rarely hear men celebrate as female; it represents, after all, an unsuitable role for a woman.

Sandilands, Eaton, and others strangely invoke a normative version of femininity, and specifically of motherhood. Eco/feminist Linda Vance provides a very different account of Mother Nature and the gendering of nature. When asked about the negative associations of the personification of nature as female and whether she would consider some other characterization, Vance (ibid., 136) "somewhat flippantly, answered that if I didn't think of nature as female, I wouldn't be able to feel such enormous pleasure in her presence. A wisecrack but with lots of truth to it. Giving nature a female identity reinforces my sense of solidarity with the non-human world." Generally, critiques of "Mother Nature" quickly reduce it to maternalism and essentialism, and seem to deny the possibility that such invocations might be informed by feminism. Vance usefully distinguishes between various gendered relationships with Mother Nature and the differing implications for men and women of invoking phrases such as "rape of the earth." Vance's female nature, then, is explicitly informed by her eco/feminist politics. Her Mother Earth differs markedly from conventional understandings in which earth as mother projects human and anthropocentric values onto nature, thus domesticating it and rendering it tame, knowable, and familiar. Vance sees unruly nature as a role model for women, unlike Sandilands and Eaton, who perceive (human) mothering as a set of practices and ideals that are being illegitimately applied to a non-gendered entity such as nature.

Likewise, in "Ecofeminist Natures from Domestication to Trauma," Sandilands (1999, 194-203) seems to imagine that any accounts of home forget all the feminist work that has pointed to the domestic sphere as a key locus of women's trauma. For many of the women I interviewed, home had clearly been a site of intense despair. Their lives and activism were about transforming the domestic and trauma. To read the "domestic" as tame and knowable is to forget fundamental feminist insights about this subject. Curiously, such analyses leave behind not the home itself, but rather feminist analyses of the home, as well as the possibility that eco/feminists might have this in mind and might reasonably take it for granted that other feminists would recognize this. Sandilands and Eaton rely on assumptions that women activists and eco/feminists are reproducing rather than challenging and transforming these metaphors. Sandilands (1999, 197) was concerned that the metaphors emphasized "nature as intimately knowable," thus tending to "obscure the Otherness of nature, the moment when nature is not female, is not human mother or sister." She (ibid., 185, 194) argued that wild and wilderness should not be confused with each other, adding in parentheses that wilderness "is a thoroughly historical construct," as if somehow the concept of wild can escape history:

> Ecofeminism has, unfortunately often, offered up a version of nature that signals a quest to ideological fullness; particularly by representing nature as female and home, many ecofeminists have produced a nature that is to appear politically through a specifically familiar women's language.
>
> But skeptical of human constructions of nature that claim to know nature fully, many ecofeminists also recognize nature as symptom, showing relations between humans and nature that cannot be expressed, relations that I will call *wild.*

Such invocations of motherhood leave domestic/wild dualisms intact and assume that mothers are knowable and tameable in a way that nature is not. This recalls Linda Williams's (1992) critique of feminist matrilineages from Chapter 3, as involving unmediated feminine communication. As I pointed out, feminist matrilineages involve work and research precisely because we do not always know our mothers. The considerable feminist literature on mother-daughter relationships would seem to belie any notion that they are transparently knowable. In fact, if Sandilands's criterion for our relationship with nature is that it must contain some sense that our knowledge is limited, that nature remains in some way unknowable and beyond us, motherhood might offer an excellent metaphor, or material-semiotic, for nature, a measure of the familiar and the strange.

Essentialism as Anthropocentrism: Mother Nature Is Not a Metaphor

Reading home or Mother Nature as metaphor relies on the separation of humans and nature, and on an implicit human exceptionalism. Making a critique of essentialism is actually an expression of an anthropocentric view of the world (cf. Gaard 2011), where humans are at the centre of things, which is where feminists who critique eco/feminism for essentialism demonstrate the limits of their theorizing. To read Mother Nature as metaphor relies on the assumption that something human (motherhood) is being applied to something non-human (nature). Yet if we refuse this distinction between human and non-human nature, such anthropocentric readings no longer make sense. Another way of putting this is that whereas cyborgs, citizens, and queers have been posed as *alternatives* to mothering nature, we may be better served to understand Mother Nature not as an essentialist opposite to the cyborg, but as much a creature of breached boundaries as the cyborg itself. Mother Nature might be perceived as a contemporary cyborg, as a Harawayean co-construction of humans and non-human nature that is about transgression and reconnection, about producing a humanity that is both more and less than kinship. Kinship here is not the social construction of the natural facts of kinship, but the breaching of the natural and social – a risky construction but no more risky than invoking vampires, cyborgs, or dogs.

Thus, it is also possible to read the other way, to read earth as mother as a move that subverts conventional meanings of motherhood, one that is less about seeing the human, and especially the female, in nature, than about the nature in humans and in women. As Terry Tempest Williams (1991) recognized above, this offers an origin story that is more complicated than a narrative of human birth. This argument also seems compatible with Greta Gaard's (1993a, 309) careful working through of the problems of "mother earth" and her insistence on the politics of eco/feminist spirituality: "Ecofeminist spirituality is ecofeminist politics is ecofeminism." I suggest that seeing earth as kin, stretching the boundaries and conventions of (human) motherhood, is a shift that could relieve the pressure on human mothers. This burden would not necessarily be projected onto nature, because from an ecocentric perspective, it is not possible to make such a demand of nature. This does not suggest that gendering relationships with nature is uncomplicated; indeed, it highlights the limitations and shortcomings of criticisms that fail to credit those who espouse such ideas with either feminism or environmentalism. We might need *more* rather than less kinship with nature, a conclusion that Haraway belatedly articulated in ending her essay "Sowing Worlds: A Seed Bag for Terraforming with Earth Others" (2013, 146) with the words "Mutter, matter, mother." She may yet write about the goddess as sacredsecular (Mani 2009) natureculture humannature.

Family Trees and Unnatural Histories of Colonialism

> We know ourselves to be made from this earth. We know this earth is made from our bodies. For we see ourselves. And we are nature. We are nature seeing nature. We are nature with a concept of nature.
>
> – SUSAN GRIFFIN, ***WOMAN AND NATURE*** (1978)

> I am protecting the rainforest.
> I am part of the rainforest protecting myself.
> I am that part of the rainforest recently emerged into thinking.
>
> – JOHN SEED, "BEYOND ANTHROPOCENTRISM" (2007)

I read the oral histories of activists as eco/feminist genealogies, as a counter to "today's genetically modified quests for self- and family knowledge" (Helmreich 2007, 231), which are so often humanist stories. A common claim is that kinship has been unsettled by new reproductive technologies, queer families, and even family tree research, all of which appear to make eco/feminist concerns about mothering nature, the planet, and children seem decidedly anachronistic. Though, of course, to read these developments in another way, there is a risk of these appearing as forms of techno-determinism – modernity/technology, whether in the guise of rapid and cheap travel, new reproductive technologies, queer families, or even the web technology that makes Mormon work on genealogy an object of gratitude for many; it can seem that eco/feminism has been left, well, holding the baby, stuck with family values. As Gibson-Graham and Roelvink (2010, 322) suggest, "responding to the challenges of the Anthropocene is not simply about humans finding a technological or normative fix that will control and restore the earth. It is about human beings being transformed by the world in which we find ourselves – or to put this in more reciprocal terms, it is about the earth's future being transformed through a living process of inter-being." Throughout this book, I suggest that genealogies are constructed, generated even, and not only by theorists attempting to chart shifts and changes in a movement. They are also constructed and reconstructed by women activists in their efforts to understand and rework their lives. These genealogies include not only families of origin, but also historical figures, famous women, and those unknown contemporary women who are friends, relatives, and co-workers, as well as trees and the earth. At a time when family is an ongoing site of struggle for feminists, constructing alternative, sustainable, and sustaining genealogies – that is, stories of who we are, where we come from, and where we want to belong – becomes a project of considerable political imperative. I have also been stretching the concept of genealogy as family history beyond its humanist boundaries, understanding it not as the reproduction of the heteronormative family or as the construction of an essentialist matriarchal connection between women, but rather as the generative opening to the possibilities of intimate relationships beyond the human(ist) family and human(ist) notions of self. In this way, genealogy is a useful tool in *denaturalizing* the family and intimate relationships, one that enables us to produce "unnatural histories of family and place."

This is not a reified sense of place as many environmentalists articulate, rather I draw on Plumwood who asked whether "discourses of place and belonging marginalise denied, dislocated and dispossessed identities, privileging 'the self-identical and well-rooted ones who have natural rights and stable homes' (Haraway [1997], 215; Plumwood [2002], 23). Is the ability to maintain access (unproblematically) to a special homeplace and to protect it, not at least partly a function of one's privilege/power in the world?" (Plumwood 2008, 140). At the same time Plumwood reflects:

> I am not of course arguing that there's necessarily anything wrong with loving a special place, or that justice demands that we each love and care for all places equally, any more than it requires one love one's child only as much as all other children and no more. But justice does require that we take some account of other children, and of our own and our child's relationship to them, perhaps even that we not aim to have our child thrive at the expense of these other children. (Plumwood 2008, 147)

I suggest that Plumwood's version of a queer ecofeminist maternal ethic of care offers something quite other than "reproductive futurism" (Edelman 2004), with her call for more multiple relationships to place rather than a "monogamous ideal" (Plumwood 2008, 147) of relationships with singular and special places. I want to use this observation of Plumwood to understand what might be at stake in a thought-provoking use of the figure of the child in the film, *Fury for the Sound* (Wine 1997). Against the outrageous clarity of Vandana Shiva's account of monocultures early in the film, and her refusal of a reproductive account of women's activism, where she insists that "movements like this [Chipko] are not born, so they never die," echoing "she goes on and on and on," the film ends with the figure of the child. The closing scenes portray slow, silent, grainy pictures of what appears to be a young white boy, among the trees in the forest, not quite playing, but certainly absorbed. Those who are prone to seeing essentialism everywhere might read this conclusion as a generic and banal statement about our need to save the planet for our children, or about the importance of childhood connections with nature, but I argue that its point is much more challenging. Rather I suggest that the film poses a question about the next generation of young men on Vancouver Island, about what life they will lead and what impact this will have on the forest. The boy in the film inhabits and is implicated in a world where technology increasingly diminishes forestry jobs for men; where unemployment has always been high for women, whose temporary jobs in the tourism service sector rely on the persistence of forest; where First Nations unemployment is staggering; where the co-management of logging and its consequences remain controversial; where the First Nations never ceded their land to the Crown; where First Nations children have been forcibly removed from their families and kept in residential schools; where the question of what alliances he will make is critical – in determining what the possibilities will be for creating more sustainable ways to live and work and support families, and communities, with the forests and creatures of the island.

Careful attention to the work of living and campaigning in Clayoquot reveals profound challenges to the maternalism and universalism that supposedly permeate ecofeminism, not only through explicit challenges to the state on the blockades and through campaigns like Brazil of the North, but just as much through everyday efforts to rework the heterosexual nuclear family, in places where, as Emberley, Perry, and others have argued, the bourgeois family has been a central site of colonial power (Emberley 2007; Perry 2001).

9

Eco/Feminism and the Changing Nature of Feminism

The Ending of Feminism? She Goes On and On and On

I set out to document Clayoquot's eco/feminist politics, hoping that I could produce an account of eco/feminism, and eco/feminist activism in particular, that could not easily be reduced to essentialism and universalism. I have offered an alternative to the tripartite possibilities of essentialism, anti-essentialism, and strategic essentialism by providing altogether different approaches to the supposed evidence of essentialism in women's environmental activism. I have retold the supposedly familiar stories of motherhood and care for children as an impetus to eco/feminist activism, of peace camps as sites of maternalist women's caring, of the planet as home, of nature as untouched wilderness, and of childhood as the site of an unmediated, innocent relationship with nature and as a naturalized origin for the adult activist self. I have produced decidedly other stories of eco/feminism, through retelling the lives and activism of women at Clayoquot. Despite the previous efforts of eco/feminist academics, it seems that an explicit commitment to either anti-essentialism or strategic essentialism has not earned eco/feminists a place at the feminist table, which they/we have so desired. Anti-essentialism has not been enough. And neither am I confident that my efforts will fare any better in stemming accusations of essentialism, universalism, and the end of feminism.

In the meantime, there have been other movements in feminism. Discussions of postfeminism, third waves, fourth waves, and generational conflict have been ongoing almost since the end of feminism was proclaimed. As Misha Kavka (2002, 29) suggests, speaking of postfeminism (although her remark is equally applicable to the end of feminism), "its definition, appropriation, or rejection has itself become a site of feminist politics." Recently, a range of critical feminist histories has revisited the dominant stories of feminism, including Wiegman (2000), Adkins (2004), Hemmings (2005, 2011), McRobbie (2009), and Walby (2011). Moreover, the last few years have witnessed the emergence of various iterations of new materialist feminisms (see Alaimo and Hekman 2008; Coole and Frost 2010). This is particularly intriguing from an eco/feminist perspective, not least because the focus is on "matter" rather than on "nature" per se.

I started out on this book motivated by the contrast between the vitality of eco/feminism and Clayoquot, and the emergence of claims of the demise of feminism. I have also been puzzled by the persistence of these accounts, not least because it seems to me that the end of feminism has never been true in any literal sense – it never actually happened. And if the claim cannot be taken literally, we must ask what it is for and what it does. Others have also queried feminism's supposed end. In her examination of the mechanisms of feminism's demise, Mary Hawkesworth (2004) reflected that feminist obituaries constitute "death by report," and Robyn Wiegman (2000, 808) suggested that we have seen "narrative proclamation" of the end of feminism. The repeated announcement of feminism's demise appears to be less an after-the-fact report and more a performative enunciation – but not one that has actually brought the end of feminism into being. It seems that the end of feminism has been a kind of persistent repetition, a refrain even, which in its very repetition belies any too literal meaning.

This cacophony of voices on the end of feminism (much of which comes from feminists, not solely the right-wing neo-liberal press) is perversely suggestive. The apparent fascination with endlessly poking the supposed corpse of feminism, perhaps to check if it still moves, if it is not quite dead yet, if there is still life, actually suggests to me the continuing vitality of feminism and in fact an ongoing interest in and passion about its fate. Despite – or because of – the seemingly endless repetition, feminists' ability to kill off feminism seems limited; regardless of our despair and/or murderous intentions, feminism lives on. This reminds me of Robyn Wiegman's (2000, 808) point that feminism is not self-identical – it cannot be collapsed into our personal histories, our own perhaps inevitable and definitely incredibly painful moments of despair about it, or our relationships with other feminists, who are sometimes our closest friends. That our own feminisms may have been sorely tested does not necessarily mean that feminism is obsolete.

Paradoxically, then, the repeated claims of feminism's demise may have worked to keep it alive, at a time when some feminists were unsure of how to proceed. The end of feminism may yet prove a generative refrain, a secular refrain for those who didn't know how to keep faith in feminism, or perhaps a sacredsecular refrain, to invoke Lata Mani's (2009) term. Yet, while reading the end of feminism against the grain, and suggesting the very repetition of this as a perverse sign of feminism's ongoing vitality, I do not want to go so far as to suggest that this counternarrative is enough, not least because there is a considerable body of evidence that demonstrates the persistence of feminism and feminist activism. While some were consigning feminism to the past, others, such as those in Clayoquot, were singing she goes on and on and on.

The end of feminism appears then as a universalizing narrative, which was never true for all feminisms, or all feminists. Perhaps certain forms of feminism did end, or were made to end. As Hawkesworth (2004, 963) suggested, the "textual accounts of death serve as allegorical signs for something else, a means of identifying a perceived danger in need of elimination, a way for a community to identify itself through those it symbolically chooses to kill." Thus, only certain kinds of feminism were ended, one of which was eco/feminism. Death was not by natural causes, as Hawkesworth discerned.

Clare Hemmings (2005, 115) has provided productive reflections on "the dominant stories that academics tell about the development of Western second wave feminism." She identified a narrative of loss, charting the decline of feminism from the heady days of the 1970s, when the second wave emerged, to the end of feminist activism by the close of the 1980s, usually killed off by post-structuralist theory. In this trajectory, feminism shifts from a politicized and unified early second wave to entry into the academy in the 1980s and to fragmentation into multiple feminisms and individual careers. The story is one of *loss* of commitment to social and political change. A second narrative is based on ongoing progress – from the naive, essentialist radical feminism of the 1970s, through black feminist critiques and the "sex wars" of the 1980s, to the sophisticated post-structuralist theory of the "difference" 1990s and beyond. This version is one of *progress* beyond falsely boundaried categories and identities.

Hemmings specified two variants of the feminist loss narrative. The first is a simple reverse story in which the essentialist 1970s are marked as good, rather than bad, as a time of unity before post-structuralist theory called into question the foundations of feminism in the category of woman and hence the possibility of feminist politics. This story relies on fixing the meanings of decades and on the supposed unity of feminist politics during these decades, as well as encouraging nostalgia for the apparent innocence of the 1970s. Hemmings (2005, 126) characterized the second loss narrative as wishing for a return to a feminism that predated post-structuralism, so that we can then "move forward" to an era that will revalue an imagined past. Despite their differing inflections, Hemmings demonstrated that both narratives rely on fixing certain feminist conversations in specific decades in order to produce an account of feminism as having changed: the 1970s are essentialist, the 1980s are dominated by critiques from black feminists and the sex wars, and the 1990s appear as the decade where a sophisticated post-structuralist feminist theory emerged. Although the loss narratives vary over the value attributed to the feminism of the 1990s, they both nonetheless leave the chronology of feminism intact. Hence, both are based on a developmental approach, where feminism moves from a preoccupation with unity and sameness, through identity and diversity, and on to difference and fragmentation.

Essentialism in Narratives of Feminism: The Essential(ist) Seventies

Of particular interest for my purpose here is Hemmings's account of how essentialism figures in these narratives, which rely on locating essentialism in the 1970s. In progress narratives, the feminism of the seventies is secured as essentialist through comparing it unfavourably with the post-structuralist feminism of the 1990s. Hemmings noted that feminist post-structuralism was required to undo the essential or "natural" conception of the self even as it simultaneously maintained the category of women, which feminism necessarily entails, already naturalizing a particular chronology of feminism.

Hemmings (2005, 121) also observed that innocence is key in these narratives. The feminism of the 1970s must be presented as being unaware of its essentialist exclusions in order to allow for, and explain, the sense of loss in nostalgic accounts of a once-unified feminism now fragmented. The location of essentialism in the naive past allows it to be cast as innocent and to be excused in the progress narrative as a slightly amusing, slightly embarrassing folly of an earlier less-developed self. In pointing this out, Hemmings has offered some resources for understanding both eco/feminism's dismissal from the feminist sisterhood for its essentialism and why an insistent anti-essentialism has not been sufficient to gain a place at the feminist table.

Essentialism also figures in feminism other ways. Whereas Haraway has argued that nature is a trope on which much turns, one can also suggest that, in feminism, much turns on the trope of essentialism, though it has been invoked to different ends at different times, usually to draw lines. Certainly, a founding gesture of feminism was the rejection of essentialism, particularly its manifestation as biological determinism. Questioning any account of women's roles as natural because biological, and so inevitable, has been vital for feminism. This was extended to encompass social determinisms too. Here essentialism marked the dividing line between feminism and patriarchy. But increasingly, essentialism was invoked *within* feminism to mark distinctions between feminisms, with the fragmentation/proliferation of feminisms. This was the case for feminisms such as socialist feminism and radical and/or cultural feminism (Jaggar 1983; Alcoff 1988; Echols 1989). Naomi Schor's (1994, vii) reflection that "the essentialism-anti-essentialism debates define 1980s feminism" alerts us to the power of essentialism in defining periodizations of feminism. As Hemmings demonstrated, essentialism has now been used to mark *temporal* boundaries between the second wave and what came after. This insight is useful for understanding eco/feminism's displacement from mainstream feminist narratives.

The Untimeliness of Eco/Feminism

It therefore appears that, not only is eco/feminism supposedly essentialist and universalist, but that it compounds its crimes, its lack of sophistication, its theoretical naïveté, through being essentialist and universalist *at the wrong time.* To elaborate – though eco/feminism's emergence is traced to the 1970s and earlier, it is more commonly located in the 1980s and 1990s, thus exceeding the necessary temporal container of the 1970s for essentialism. Given this, eco/feminism can never even be a site of innocence, nostalgia, or loss. By the late 1980s and early 1990s, eco/feminists should have known better than to be essentialist; eco/feminism emerges as out of step with mainstream accounts of feminism. Eco/feminists' insistence on re-opening the supposedly closed questions of women and nature disrupts efforts to depict essentialism as safely confined to the past, and therefore eco/feminism challenges the progress narrative of feminism. But eco/feminism's challenge – threat – was that, in offering a feminist re-engagement with nature, it might fatally undermine feminism's efforts to secure a place in the academy.

Hemmings recognized that the so-called problem of essentialism was often resolved theoretically, by relying on a simplistic opposition of essentialism and anti-essentialism, and posing essentialism as an impasse that feminists must overcome, usually through a theoretical process of "linear displacement." Thus, if the progress narrative is to be secured, eco/feminism must be situated in the past, left behind in the seventies, or even better, not remembered at all and subject(ed) to a collective amnesia.

In seeking alternative approaches, Hemmings (2005, 130) asked, "How might feminist theory generate a proliferation of stories about its recent past that more accurately reflect the diversity of perspectives within (or outside) its orbit? How might we reform the relationship between feminism's constituent parts to allow what are currently phantom presences to take shape? Can we do feminist theory differently?" In posing these questions, she echoes Ahmed's (1998, 91) insistence that there cannot be a teleological settlement: "Essentialism is not a conceptual horizon that can simply be transcended." As Ahmed (ibid.) elaborated, "The 'problem' of essentialism is not one which can be resolved once and for all, but rather demonstrates the importance of a continual reflexivity over how the borders which sustain conceptual entities such as 'woman' and 'women' are constituted through acts of exclusion or othering."

Thus, Hemmings demonstrates why efforts by eco/feminists to produce eco/feminism as anti-essentialist, or only strategically essentialist, have had limited success. No variety of anti-essentialism or strategic essentialism will *ever* be sufficient, because they do not address the place of essentialism in narratives of feminism. As long as such narratives remain dominant, eco/feminism will always be out of step. And as long as conflicts in feminism are resolved through narrative, often by fixing undesirable elements in the past and then moving on from them, eco/feminism will almost always be left behind.

New Material Feminisms?

While eco/feminists have been caught up in the constraints of essentialism, other feminists have worked under the rubric of "matter" or "materialism" rather than nature as such, constructing new material feminisms and neomaterialisms in the process. Of course, Donna Haraway always did write about nature and matter, but more recent contributions include Karen Barad (2003, 2007), Elizabeth Grosz (2004, 2005), Elizabeth Wilson (2004, 2010), Vicky Kirby (2008), Stacy Alaimo and Susan Hekman (2008b), and

Diana Coole and Samantha Frost (2010). Intriguingly, these writers appear to proceed without the taint of essentialism. How they manage this is an interesting puzzle in itself, not least given eco/feminists' struggles to secure a future for themselves.

I focus on two particular points with respect to work on new feminist materialisms – first on the question of the new, and second, and relatedly, on which "matter" is "made to matter." I am not alone in stumbling over the "new" of new material feminisms; Sara Ahmed (2008) and Maureen McNeil (2010) have also raised this issue. Both focus particularly on the work of Elizabeth Grosz, maintaining that new materialists forego a thorough engagement with existing work in feminist biology, which they also link with the rejection of critique. Ahmed is concerned with the "founding gestures" of the new materialists, specifically their tendency to construct feminism as anti-biological. Ahmed's (2008, 24) intent was not to dismiss the new materialisms but rather to attend to the possible "problematic consequences for our understanding of the genealogy of feminist thought." She (ibid., 36) reflected that

> when we describe what it is that we do, when we consider how it is that we arrive at the grounds we inhabit, we need to appreciate the feminist work that comes before us, in all its complexity. We don't always have to make a return to earlier feminist work, but if we represent that work as being this or that, then we need to make that return. Such a return would be ethical: we should avoid establishing a new terrain by clearing the ground of what has come before us. And we might not be quite so willing to deposit our hope in the category of "the new."

In a later essay that appeared in *New Materialisms,* Ahmed (2010, 234) repeated,

> In thinking the dynamism of matter, this essay joins a body of scholarship that has been called by the editors of this volume a "critical materialism." I would nonetheless resist calling my own contribution a "new" materialism inasmuch as my own work draws on, and is indebted to, earlier feminist engagements with phenomenology that were undertaken during the period of "the cultural turn." These phenomenological engagements belie the claim made by some recent materialist critics to the effect that, during this period, matter was the only thing that did not matter.

Grosz and others assert feminism's need to address biology, although as Ahmed observed, this assertion is made possible only by "forgetting" that a significant amount of feminist work has already done so, and I would add, that the intense critique of essentialism often made it difficult to speak of biology. This convergence of a rejection of critique with a renewed focus on nature and biology is indeed provocative in the context of eco/feminism and is suggestive of one of the ways in which such theorists avoid the label of essentialism or even anti-essentialism. Ahmed and McNeil both appear to share an understanding of critique as being a close and generous engagement with that which is useful in the work of other feminists, including (for Ahmed explicitly) earlier feminists, but it is not clear that this is Grosz's understanding. Thus, for both Ahmed (2008) and McNeil (2010), the move toward an abandonment of critique is a concern, particularly in the work of Grosz though also in that of new materialists more broadly. As McNeil (ibid., 433) suggests, it seems impossible completely to abandon critique, and the god-trick of seeing from nowhere has long been a focus of feminist concern. Thus, the supposedly affirmative approach of the new materialists appears misguided, particularly if we take seriously Hemmings's (2011) account of the politics of citation – that lack of citation in effect disestablishes other histories.

Interestingly, Ahmed's 2008 essay provoked strong defences of new materialisms, including of Grosz's work, by Iris van der Tuin (2008, 2009), Noela Davis (2009), and Myra Hird (2009). However, this work does not appear to address Ahmed's central arguments about time, history, and genealogy, as well as ethics, accountability, and responsibility. In their very rebuttal – that 1970s feminist work on biology is not the same as that of the new materialisms and that, implicitly, it properly belongs in the past – Davis, Hird, and van der Tuin appear to restate their insistence on the newness of the new materialisms, and thus do not appear to engage with the spirit of Ahmed's intervention. For instance, Noela Davis (2009, 79n6) insisted in an endnote that "arguments against biological determinism were important in their time but, while they paved the way for later feminist work, they underestimated or dismissed the importance of biology to our sociality," a comment that both homogenizes the feminist seventies and reinstates a particular version of feminist history. The question that Ahmed posed was not so much what is new about the new materialisms, but rather *what is at stake* in claims to the new and *how* this new is established. Ahmed was concerned with what is made not to matter and why, in the process of making something new. It is not clear that these questions have been adequately addressed by those professing versions of the new materialisms, but they remain important in understanding what new materialisms might have to offer feminism more broadly. Otherwise, new materialisms seem to be little more than a variant of Hemmings's progress narrative of feminism.

Davis, Hird, and van der Tuin insist on the importance of "engagement" rather than "critique," here meaning a feminist engagement with science, as opposed to a critique of it, which is how 1970s feminism seems to be characterized. Spivak's (1994, 157) comment on critique bears repeating here: "The critique of essentialism should not be seen as being critical in the colloquial, Anglo-American sense of being adversely inclined, but as a critique in the very strong European philosophical sense, that is to say, as an acknowledgement of its usefulness." The lessons of the possibilities of deconstruction appear to have been lost. At the same time, the attempt to move from the supposed limits of critique to an affirmative feminism without comprehensively addressing the damaging consequences of critique – and the critique of critique – seems risky. This is particularly the case if critique is replaced with an affirmative approach that obfuscates its own allegiances and inevitable disavowals.

It is not clear to me that eco/feminism would fare any better with "engagement" than with "critique." It is worth noting that both Ahmed and McNeil, as well as the new materialists themselves, are largely silent about eco/feminism, a point that is also made by Richard Twine (2010, 402): "Whilst Ahmed criticizes the new materialists for inadequately acknowledging feminist work on the biological, neither she nor most of the new materialists acknowledge ecofeminist scholarship." Twine (ibid.) asserts that "the emergence of a feminist new materialism ought to usher in a renewed conversation between feminism and ecofeminism due to shared interests." Twine's important article focuses specifically on the growing attentiveness to research on non-human animals and contends that it is no longer viable for feminists to ignore the non-human. Interestingly, given the importance of temporality in narratives of eco/feminism, he suggests that humanist disgust rather than eco/feminism should be seen as out of date. Yet as Twine himself notes, the new materialists rarely mention eco/feminism. It is Haraway (2008), for all the controversy over her account of companion species (see Weisberg 2009 for an overview), who acknowledged that eco/feminists worked on animals long before many became interested and who also continues to insist on the importance of eco/feminist genealogies. Given the new materialists' commitment to the new, and their concomitant partial histories, and insistence that feminist materialisms are quite other than the feminist biology of the 1970s, it is difficult to see how they would recognize shared interests or shared natures.

For instance, juxtaposing Grosz's lack of acknowledgment of eco/feminism with her commitment to what she terms "positivities" is certainly revealing. In *Time Travels: Feminism, Nature, Power,* Grosz (2005, 2) insisted, "Rather than undertake the expected path of political and philosophical analysis, in which a thinker's position is subjected to rigorous criticism and its errors, contradictions and points of weakness are singled out or overcome, I am more concerned with seeking out positivities, crucial concepts, insights on what is of value in the texts and positions being investigated." McNeil contrasted this comment with Grosz's dismissive account of feminist work on Darwin. For my part, I want to juxtapose it with her disavowal of eco/feminism and any need to *engage* with it – and this in a book that is subtitled *Feminism, Nature, Power* – "With the exception of the ecology movement, with its eco-feminist and eco-philosophy offshoots, with which I am loath to be identified, virtually all forms of contemporary political and social analysis continue this tradition of ignorance of, indeed contempt for, the natural, which today remains identified with either passivity or inertia" (ibid., 34). Here Grosz spectacularly refuses to trace feminist histories through ecofeminism. Indeed, eco/feminism is severed from any connection with feminism and relegated to an arborescent offshoot of environmentalism. This unfortunate and uninformed dismissal is neither critique, nor engagement, nor affirmation. In this way, Grosz authorizes her own work and perhaps that of other new materialists as the rightful interlocutors on the matter of nature, disavowing others in the process and leaving the possibilities of shared interests decidedly shaky. Her loathing resonates with Twine's (2010, 402) reference to disgust – he reflects that "it would be a shame if disgust were to get in the way of conversation" between eco/feminists and other feminists, but without a more meaningful commitment to engagement, or even critique, in Spivak's sense, it is difficult to see how the conversation will come about.

The collection edited by Stacy Alaimo and Susan Hekman, called simply *Material Feminisms* (2008b), appears to open up more possibilities for dialogue. Perhaps significantly, the word "new" does not appear in the title, and "material" modifies "feminisms," rather than vice versa. Furthermore, the book explicitly names eco/feminism alongside new feminist materialisms and creates the possibility of the traffic that Twine anticipated, rather than dismissing it, like Grosz. There is much to commend in this collection, including the way in which chapters by Barad, Wilson, and Kirby jostle alongside essays by eco/feminists. Although none of the chapter titles refer to eco/feminism per se, Alaimo and Hekman's (2008a, 10) introductory essay states the thesis "that feminist theory is at an impasse caused by the contemporary linguistic turn in feminist thought," and it presents the anthology as "a first step in not just articulating these theories but integrating them into what amounts to a new paradigm for feminist thought." They gesture to alternatives to the dominance of the linguistic turn as, for instance, in the formulation of the "material-discursive" in the chapter by Haraway. Even so, Haraway's essay was first published in 1992 and thus has been around for some time. I present this as just one example of the long history in feminism of thinking the material and the discursive together. This indicates that the problem may not be the linguistic turn per se, but rather the stories we tell about feminism, the investment in the new, in teleological narratives of progress, and a failure to attend carefully to what is at stake in the telling of stories and in the nature of feminist history.

(Re)reading a classic eco/feminist work such as Susan Griffin's *Woman and Nature: The Roaring inside Her* (1978) in conjunction with some of the new materialisms might be instructive. Stacy Alaimo (2000, 180-82) did revisit Griffin's text, and though she initially found much of interest, she ultimately read it as reverting to dualisms. Yet it could be invoked to question the supposed newness of the focus on matter. A rather different opening out would have been possible had Griffin's text been read alongside, say, Vicky Kirby's essay "Natural Convers(at)ions: Or, What If Culture Was Really Nature All Along?" (2008). Griffin's *Woman and Nature* consists of three "books," with the first devoted to "Matter." In the third, entitled "Matter Revisited," Griffin (1978, 226, emphasis in original) wrote, "*We know ourselves to be made from this earth. We know this earth is made from our bodies. For we see ourselves. And we are nature. We are nature seeing nature. We are nature with a concept of nature. Nature weeping. Nature speaking of nature to nature.*" What might Kirby's meditations on natural conversations bring to our understanding of Griffin's book – and vice versa? How has Griffin's radical feminism been relegated to Haraway's "organic holism," while, say, Daphne Marlatt, who cites Griffin in the epigraph to *Ana Historic*, has often been understood as a post-structuralist feminist?

How is Griffin's work read as essentialism, rather than seen as offering an account of interconnection?

Much of the new feminist materialisms are oriented to, or more precisely against, the discursive or "linguistic" turn, rather than around histories of nature in feminism. McNeil noted this move in Grosz's account of some feminists' focus on "cultural construction at the expense of natural production" (quoted in McNeil 2010, 433); it is also evident in Alaimo and Hekman (2008a), and most particularly in Karen Barad's (2003, 801) founding gesture for her post-humanist performativities:

> Language has been granted too much power. The linguistic turn, the semiotic turn, the interpretative turn, the cultural turn: it seems that at every turn lately every "thing" – even materiality – is turned into a matter of language or some other form of cultural representation. The ubiquitous puns on "matter" do not, alas, mark a rethinking of the key concepts (materiality and signification) and the relationship between them. Rather, it seems to be symptomatic of the extent to which matters of "fact" (so to speak) have been replaced with matters of signification (no scare quotes here). Language matters. Discourse matters. Culture matters. There is an important sense in which the only thing that does not seem to matter anymore is matter.

It is difficult not to read this as deliberately provocative; after all, the rest of Barad's article takes a much more nuanced approach to language, discourse, and matter, which is belied by this opening salvo. Furthermore, Barad concluded her book *Meeting the Universe Halfway* (2007, 394) with reflections on consequentiality, responsibility and accountability, and exclusions – arguing that "accountability and responsibility must be thought in terms of what matters and what is excluded from mattering." Yet it must be said that not all new feminist materialists are attuned to these issues; for many, not all matter comes to matter. Crudely, new matter may be made to matter more than old matter.

This is not the case for all materialists. For example, in "Materialist Returns: Practising Cultural Geography in and for a More-Than-Human World," Sarah Whatmore (2006) offered a careful account of the current turn to matter, precisely because of her attention to historicity and temporality. Her essay demonstrates both a historical and a geographical imaginary. Whatmore (ibid., 601) proposed that "what is new (as in different) about the something/happening in cultural geography is a product of repetition – turning seemingly familiar matters over and over, like the pebbles on a beach – rather than a product of a sudden encounter or violent rupture." In using the language of turns and returns to materiality, Whatmore (ibid., 600) also draws on her early work as a rural geographer to remind readers that her "obdurately earthy interests in cultivation and property, growing and eating" were not always fashionable. Though focused on cultural geography, her remarks could also apply more generally to the turn to matter and could offer some useful pointers to those who proclaim "new feminist materialisms." Accounts such as that offered by Whatmore can be usefully understood as "knotted pastpresents" (King 2010). Whatmore, King, Ahmed, Hemmings, and numerous other feminists all have many interests, new, old, and returned to, including matter and discourse. Against the new materialists' rather particular version of (feminist) history, such feminisms appear to be the "modest witnesses" of our times.

The new materialisms serve as a potent reminder that though nature and essentialism are often aggregated in feminist theory, this is not always the case. Ahmed (2008, 36) reminds us not to lose sight of "how matter matters in different ways, for different feminisms, over time," and the same might be said of nature. Ahmed's attention to the genealogy of feminist thought is key here, and it returns me to Sturgeon. Sturgeon's genealogical critique of typologies may remain pertinent – after all, post-millennial feminist theorists do not necessarily write so much about activism or spirituality or feminists of colour, raising the question of what (else) they might have had to jettison to avoid the accusation of essentialism.[1] Paying attention to which biologies, natures, and matter are remembered, and which forgotten, may be crucial.

In their insistence on the new, the new feminist materialists risk perpetuating a teleological progress narrative, not least because in focusing on matter rather than nature, they can easily forget that the problem with nature is not actually nature per se, but rather the (mis)conception that it is fixed and deterministic and therefore not a site for social and political change. In the refusal of critique for positivities and other affirmative stances, there is also a risk of a lack of accountability and responsibility. In short, new feminist materialisms seem as unable to address conflict and the co-existence of difference as many other feminisms; this is certainly not a new challenge for feminism. Michelle Bastian offers a useful intervention here. Her work on community and time demonstrates how linear time is used in the project of inclusion and exclusion in community. Bastian (2011,

152) turns to Gloria Anzaldua for one feminist engagement with this subject, stating that Anzaldua's writing "challenges the tendency to separate differences out across time and instead utilises a notion of simultaneity that seeks to enable the recognition of difference within the 'same' moment of time." In noting that a common approach to dealing with political questions is to arrange them hierarchically in order of importance and address them sequentially, Bastian (ibid., 155) asserts that "the attempt to deal with 'one thing at a time' provides a model for how groups should arrange competing values and meanings, in that the diversity of demands for response cannot be responded to 'at once,' but must be ordered and arranged in terms of importance. Differences are no longer simultaneous but are separated out and spread across time." As Bastian (ibid., 158) suggests, "To be coeval (i.e., to live in the same time with another) is to recognise the multiple lines of time and of history that operate within the present, in ways that are not fully commensurable." She (ibid., 162) goes on to propose that "one way of tackling both the limits imposed by such histories is to develop an account of time which will enable contradictory histories and contradictory ways of acting to share the same time, to be coeval with each other, rather than be divided from each other as they are within linear accounts of time." Her version of the untimeliness of feminist theory differs markedly from that of Grosz (2010) and does not rely on exclusions or on separating out feminisms over time, offering a way of seeing the persistence of eco/feminism in the '80s and '90s and beyond the supposed essentialism of the '70s.

Re-Visioning the Changing Nature of Eco/Feminism through Clayoquot: What Can Be Salvaged?

> I've been working on the sense of salvage as "something extracted (as from wreckage, ruins, or rubbish) as valuable," and it seems to me that this kind of salvage involves revaluing the way that we've just been talking about, re-visioning (I do use Adrienne Rich's term for it), and neither of those acts seems to participate in sophistry. In fact, they're essential for a woman's coming to speak out of her hitherto absent body, absent desire, and muted voice within a public context. To do this she must valorize the ground out of which she speaks. Which means working for really drastic change in our culture.
>
> – DAPHNE MARLATT, 1991, QUOTED IN BRENDA CARR, "BETWEEN CONTINUITY AND DIFFERENCE"

Clayoquot offers an exemplary site through which to examine ongoing challenges in feminism, particularly the place of essentialism and how essentialism and nature/culture dualisms figure in the narration of the end of feminism. Despite widespread claims that feminist activism was obsolete, it persisted and flourished at Clayoquot. Clayoquot confounds the too literal meaning of the end of feminism and demands that we re-vision our histories of recent feminism. Eco/feminist activism in Clayoquot emerges not as a throwback to the disavowed efforts of the 1970s and 1980s, not as a site for the repetition of essentialism or anti-essentialism, but as a site where we can observe feminism and eco/feminism in the making – and unmaking. Clayoquot offers us an opportunity to imagine the eco/feminist past somewhat differently: as a series of ongoing conversations rather than a process of linear displacement, as a site of ongoing reflexivity regarding the construction of boundaries between conceptual entities such as "women" and "nature," and as a site through which we might re-vision feminist histories that insist, not so much on anti-essentialism, but on the inevitable essentialism of any articulation of women and nature. Essentialism has become essential to (narrating) eco/feminism; the question is what possibilities have been foreclosed in this process? Yet, once the "ism" is removed from essentialism, eco/feminism might serve to remind us of the continuing need to explore the question of *what is essential to feminism* – all of which suggests that eco/feminism may yet have much to promise.

Appendix

Interviews

Name	Location	Date
Abbey, Irene	Vancouver	4 July and 9 September 1996
Back, Kim	Denman Island	16 July 1996
Bassingthwaite, Linda	Gabriola Island	11 July 1996
Bate, Jan	Tofino	27 September 1996
Bell, Louise	Denman Island	15 July 1996
Browne, Greg	Denman Island	7 September 1996
Fawkes, Jane	Denman Island	7 September 1996
Fireweed	Denman Island	14 July 1996
French, Jacinta	Vancouver	1 July 1996
Gel, Atty	Vancouver	3 July 1996
Hamburger, Jay	Vancouver	23 September 1996
Hayvice, Christine	Vancouver	10 July 1998
Kagis, Dana	Tofino	13 August 1996
Kanetsuka, Kami	Vancouver	3 July 1996
Krawczyk, Betty	Tofino	22 July 1996
Langer, Valerie	Tofino	14 August 1996
Leigh, Miriam	Denman Island	7 September 1996
Lillico, Norleen	Tofino	13 August 1996
Lowther, Chris	Tofino	29 September 1996

►

Name	Location	Date
McLaren, Jean	Gabriola Island	10 July 1996
Morrell, Mike	Denman Island	6 September 1996
Mychajlowycz, Maryjka	Tofino	28 September 1996
Ormond-Twiss, Inessa	Vancouver	30 June 1996
Paone, Sergio	Tofino	27 September 1996
Pope, Wendy	Denman Island	17 July 1996
Rogers, Sophie (pseudonym)	Vancouver	15 September 1996
Simpson, Amy	Vancouver	16 September 1996
Simpson, Síle	Denman Island	15 July 1996
Thompson, Judith	Vancouver	16 September 1996
Wilson, Diana	Vancouver	6 July 1996

Notes

Preface: "She Goes On and On and On"

1 For the song lyrics and the extract from *Spare Rib* 142 (1984): 27-28, see Danish Peace Academy, http://www.fredsakademiet.dk/abase/sange/greenham/song4.htm; for the Michael Marriott sculpture at Greenham Common, see Greenham Common Women's Peace Camp Commemorative and Historic Site, http://www.greenhamwpc.org.uk/historic.htm.

2 Many thanks go to John Othick and others at the Centre of Canadian Studies, Queen's University Belfast, for their early support for this project. For more on the Queen's University Centre of Canadian Studies, see http://www.qub.ac.uk/schools/CentreofCanadianStudies.

Chapter 1: Rethinking Eco/Feminism through Clayoquot Sound

1 This point has also been resoundingly made by Warren Magnusson and Karena Shaw in their thoughtful and thought-provoking collection *A Political Space: Reading the Global through Clayoquot Sound* (2003c, viii), which avoids easy answers to the question of what, or where, Clayoquot is. The collection offers an excellent guide to the politics of Clayoquot, as does the associated website of the Clayoquot Project. See Clayoquot Sound Research Group, "A Political Space: Reading the Global through Clayoquot Sound," 2002, http://web.uvic.ca/clayoquot/clayoquotWorkshop.html.

2 On environmental politics, see Braun (2002), Magnusson and Shaw (2003b), and Castree (2005). For useful anthologies, see Berman et al. (1994) and Breen-Needham et al. (1994). For dissertations, see Berman (1995), Boucher (1997), and Moore (2002). See also Hessing, Raglon, and Sandilands (2005).

3 For the extensive and thorough archive of documentation relating to the campaign over the forests of Clayoquot, see http://web.uvic.ca/clayoquot/clayoquotWorkshop.html.

4 As well as the FOCS, the Western Canada Wilderness Committee (now the Wilderness Committee), Greenpeace, and the Women's Environmental Network were involved in the campaign. It is worth stressing that although the organizations differed in their focus and strategies, there were also many connections between them. For instance, in her interview Valerie Langer mentioned the importance of networks of women in environmental organizations across British Columbia, and the support and mentoring that she gained through these informal networks.

Chapter 2: Eco/Feminist Genealogies

1 Though note Sandilands's comment in Gaard (2011, 45n6): "I have identified myself as an ecofeminist consistently since I first read the term."

2 The subheading of this section is adapted from Cuomo (1998, 5).

3 Titled "Ecofeminism: A Practical Environmental Philosophy for the 21st Century," the conference was held at the University of Montana, Missoula, 2-5 April 1998. See http://www.cep.unt.edu/news/mt.html.

4 Although she did publish her first monograph, *Ecological Politics,* in 1998. Perhaps by that time, monographs did do justice to eco/feminism as it had become. More recently, Gaard has written a work of creative non-fiction, *The Nature of Home: Taking Root in a Place* (2007), which makes no explicit mention of eco/feminism until it reaches the closing acknowledgments.

5 The conference was "Praxis/Nexus: Feminist Methodology, Theory, Community," at the University of Victoria, Victoria, BC, 18-20 January 1996.

6 The Brazil of the North campaign and this quote have been credited to the late Colleen McCrory, who set up the Valhalla Wilderness Society. See "Colleen McCrory," The Goldman Prize, n.d., http://www.goldmanprize.org/taxonomy/term/66?page=1; and also "Canada: The Brazil of the North," n.d., http://web.uvic.ca/clayoquot/files/volume1/III.D.5.pdf.

Chapter 3: Eco/Feminism and the Question of Nature

1 Though for some it may seem odd to bring Haraway and Starhawk together, I am not alone in refusing to read the cyborg and goddess as oppositional figures. See Haran (2003, especially Chapter 6 on magic, technology, and spirituality; Haran 2010, 2014) and the work of Maria Puig de la Bellacasa, Isabelle Stengers, and Debi Withers. As Noël Sturgeon (1997, 24) notes, "A name that can usefully if partially describe the work of Donna Haraway and Mary Daly, Alice Walker and Rachel Carson, Starhawk and Vandana Shiva, ecofeminism is a shifting theoretical and political location that can be defined to serve various intentions." Also, as I discuss later in the chapter, a number of eco/feminists have taken up the cyborg; as well as Sturgeon (ibid.), see Sandilands (1997, 1999) and Cuomo (1998).

2 Haraway does acknowledge Starhawk's influence in the spiral dance and activism, in an interview with Jeffrey Williams published in the *Minnesota Review* (Williams 2009, 138-39).

3 That said, the significant point of difference between Haraway and some eco/feminists might not be goddesses or cyborgs at all (or the matter of meat-eating or agility training), which is in itself revealing, but the matter of (human) population. In her interview with Joseph Schneider (Schneider 2005), Haraway referred to the earth as having a fixed carrying capacity and suggested that *not* having children might constitute an environmental good. Such definitive statements are hardly characteristic of her writing, which rarely suggests a correct mode of action or being in the world and more usually works to disrupt certain taken-for-granted ideas of feminism or environmentalism.
4 See The Fifth Sacred Thing, http://www.thefifthsacredthing.com/, for more details on the film. I am again indebted to Joan Haran for her research on the making of the film (Haran 2013).
5 See Puig de la Bellacasa (2010) on the permaculture-based Earth-Activist Training workshops that Starhawk now runs, arguably a development of witchcamp.
6 Haraway was not able to contribute to the book. For details of the workshop and summaries of the discussions, see "A Political Space: Reading the Global through Clayoquot Sound," 2002, http://web.uvic.ca/clayoquot/clayoquotWorkshop.html.
7 For Carson's letters, which have now been published, see Freeman (1995).
8 I draw on Mary Eagleton's account of feminist matrilineages in *Feminist Literary Criticism* (1996, 38), which cites Trinh T. Minh-ha's (1989) discussion of women's roles as storytellers, as keepers and transmitters of stories. Additionally, though I turn to Virginia Woolf here for her theory and methodology, it also seems important to acknowledge her commitment to pacifism and her opposition to war, most notably in *Three Guineas.* Thanks to Joan Haran for reminding me of this.
9 For more on this point, see McMillan (2004, 115).

Chapter 4: Clayoquot Histories

1 As McLaren and Brown (1993, 68) explain, "Supernatural B.C." was a slogan employed by the Social Credit government to promote tourism while it simultaneously gave away forests to foreign companies. Sealand was a Victoria aquarium that held orcas. For more on the Raging Grannies, see Chapter 6.
2 The subheading for this section is taken from Douglas MacKay's *Empire of Wood* (1982), the official history of MacMillan Bloedel, whose title is possibly more telling than was intended.
3 Daniel Clayton (1996, 98n5) wrote that "the Native groups of the west coast of Vancouver Island took Nuu-chah-nulth as their collective name in 1978. Before that Europeans called them 'Nootkans.' The Native groups of this region did not identify themselves collectively in 1778." Throughout his book, he put Nuu-chah-nulth and Nootka in quotation marks to highlight processes of mapping and renaming that were central to the colonization of Native peoples and land.
4 The Nuu-chah-nulth may in effect be settlers too, though Chief Earl Maquinna George (2003) disputes the claim that they came to Clayoquot via the Bering Strait land bridge.
5 On the quest for Eden in BC writing, see Pritchard (1992).

6 See, for example, Clayton (1996) for First Nations accounts of Cook's arrival. See also Louis Miranda and Philip Joe, "How the Squamish Remember George Vancouver" (1993) as well as George (2003). For a collection which focuses on the broader impact of indigenous thought on public discourse and policy development in Canada, see Timpson (2009). It is also worth noting that Paul Tennant (1990, 3) wrote that "whites have consistently underestimated the original aboriginal population of BC. The low number resulted in part from the scarcity of archaeological research, but it served as well the implicit purpose of demeaning aboriginal claims and buttressing white myths. In the 1970s and 1980s the estimates of the pre-contact population rose steadily. Today the estimate is between 300,000 and 400,000 for all of BC."

7 Ministry of Forests, cited in Friends of Clayoquot Sound, *Newsletter,* Spring 1996, 1.

8 I am indebted here to Ken Drushka's *H.R.: A Biography of H.R. MacMillan* (1995) for details of MacMillan's life and for quotations from his writing. The interpretations, however, are my own.

9 The subheading of this section is based on Marlatt (1990, 88): "hystery. the excision of women (who do not act but are acted upon)."

10 For more information, see Pritchard (1992), though Pritchard is more concerned with the representation of history than with considering novelists themselves as historians, actively engaged in the writing and rewriting of BC history.

11 Kogawa participated in the Rally to Save Clayoquot Sound, held at Vancouver's Orpheum Theatre on 7 November 1993, and is published in Breen-Needham et al. (1994).

12 *Ana Historic* has been described as a feminist rewriting of Martin Allerdale Grainger's 1908 novel *Woodsmen of the West,* a realist account of life in BC logging camps at the time. It also provides an interesting contrast with Kate Pullinger's (re)writing of Vancouver Island history in *The Last Time I Saw Jane,* which imagines the life of Governor James Douglas and his Metis wife, Amelia Connolly. The two books constitute differing attempts to rewrite the history of women of the region. Reweaving the narratives of the early explorers/colonizers in fiction and poetry is common in BC writing; examples include George Bowering's *Burning Water* (on Captain George Vancouver), Marilyn Bowering's *To All Appearances a Lady,* Lionel Kearns's *Convergences* (on Captain James Cook), and Ethel Wilson's *Love and Salt Water.*

13 See Lugrin (1928, 146-60) and Lay (1980) for more on the bride ships. On white women and imperialism, see Chaudhuri and Strobel (1992), Pratt (1992), Ware (1992), Burton (1994, 2001), McClintock (1995), and Midgley (1998).

14 See Friends of Clayoquot Sound, *Newsletter,* Summer 2000, 2; Friends of Clayoquot Sound, *Newsletter,* Spring 2001, 6-7.

15 The Tla-o-qui-aht First Nation was protesting the transfer of airport land to the regional district. Friends of Clayoquot Sound, *Newsletter,* Summer 2000, 1.

Chapter 5: "It Was Like a War Zone"

1 I don't mean to suggest that women-only or lesbian separatist activism is inherently essentialist. See Roseneil (2000) and Sandilands (2002).

2 McLaren's *Spirits Rising* (1994, 76) gives the full text of the welcome handout.
3 McLaren (1994) provides an account of some of the training at the camp.
4 McLaren (1994, 38, 26) briefly described this episode as well as a further one in which a man came into the camp and approached a young girl as she was changing clothes in her tent after a swim – he left when some other people came by the tent.
5 The camp had a support group for those who had been sexually abused, which met regularly, usually several times a week throughout the summer.
6 This event is also described by Chris Lowther (1997, 71).
7 This is also suggested by McLaren (1994, 14), who mentioned a boy of fifteen who spent the summer at the camp but who felt that some of the older men discounted him as "only a kid" and did not listen to his ideas.

Chapter 6: Mothers, Grandmothers, and Other Queers in Eco/Feminist Activism

Thanks to Annis May Timpson for suggestions on the title of this chapter.
1 For example, see Merchant (1981, 1996), Freudenberg and Zaltzberg (1984), Rosenberg (1987), Hamilton (1990), Nelson (1990), DiChiro (1992), Krauss (1993), Seager (1993), Brown and Ferguson (1995), T. Kaplan (1997), and Gibbs (1998).
2 To this end, feminists have produced a range of historical, anthropological, and sociological research to demonstrate the contingency of the modern family; further, feminist science fiction and utopian novels have envisioned different family and kinship arrangements. For examples, see the collections Thorne and Yalom (1982) and Trebilcot (1984). Marge Piercy's *Woman on the Edge of Time* (London: Women's Press, 1970) remains a popular example of feminist science fiction, which takes up Firestone's "demands and speculations" (thanks to Joan Haran for this).
3 Ruddick went to considerable lengths to acknowledge women's involvement in war and acts of violence. Cynthia Enloe (1983) has also written on women's various involvements and implication in militarization.
4 Among the many critics of maternalist peace politics are Dietz (1985, 1987), Spelman (1988), Collins (1990), and Bailey (1994). Critics of maternalism in eco/feminism include Biehl (1991), Roach (1991), Eaton (1997), Sturgeon (1997), and Sandilands (1999).
5 For feminist works that voice concerns about maternalism, see Segal (1987), Biehl (1991), Agarwal (1992), Jackson (1993a, 1993b, 1995), and Seager (1993). For eco/feminist works, see Warren (1987, 1990), Lahar (1991), Roach (1991), DiChiro (1992), Buege (1994), Davion (1994), Sturgeon (1997), and Sandilands (1999).
6 See also Merchant (1981), Freudenberg and Zaltzberg (1984), Rosenberg (1987), Hamilton (1990), Nelson (1990), DiChiro (1992), Krauss (1993), Seager (1993), Gibbs (1997), and T. Kaplan (1997).
7 Note that "spouses" become "husbands" in the quote.
8 See Brown and Ferguson (1995, 163), whose citations include Krauss (1993). See also Gibbs (1982), Freudenberg and Zaltzberg (1984), Rosenberg (1987), DiChiro (1992), and T. Kaplan (1997).
9 Although Sara Ruddick's *Maternal Thinking* (1989, 225-34) barely mentions grandmothers, it does briefly discuss their involvement in women's resistance to military

dictatorship in Chile and Argentina, particularly their work on behalf of the disappeared. In this case, mothers and grandmothers remember and trace their disappeared children, an effort that extends beyond their own children to all disappeared children.

10 See Elizabeth Spelman's *Inessential Woman* (1988) for an account of the additive process, or the "ampersand problem," in feminism. Hill Collins (1990) and Bailey (1994) discuss the topic with respect to issues of motherhood and race.

11 Jane Fawkes and her family were not the only ones to move to Denman in the wake of the Clayoquot campaign; Greg Browne, Fireweed, and Síle Simpson also moved to the island between 1993 and 1996.

Chapter 7: Romanticizing the (Gendered) Nature of Childhood?

1 For an early example, see Shulamith Firestone's *The Dialectic of Sex: The Case for Feminist Revolution* (1970, 72); in a chapter titled "Down with Childhood," it argued that childhood was an oppressive state for "younger people" and that feminist revolutionaries must eliminate "the very conditions of femininity and childhood themselves that are now conducive to this alliance of the oppressed, clearing the way for a fully human condition."

2 See Jones (1997, 1999).

3 See Kurth-Schai (1997) for an eco/feminist account of instances of children's environmental activism; on children's activism more generally see, for example, James (1990) and Sibley (1991). See also Ward (1978).

Chapter 8: Unnatural Histories

1 See Woods et al. (2013) for one approach to a more complicated account of arborescence and rhizomes in social movements.

Chapter 9: Eco/Feminism and the Changing Nature of Feminism

1 And, as Ahmed (2008, 36) notes, similar gestures also manifest in recent work in critical race studies.

References

Acker, A., and B. Brightwell. 2004. *Off Our Rockers and into Trouble.* Victoria: Touchwood.

Adams, C. 1991. *The Sexual Politics of Meat.* New York: Continuum.

–, ed. 1994a. *Ecofeminism and the Sacred.* New York: Continuum.

–. 1994b. *Neither Man nor Beast: Feminism and the Defense of Animals.* New York: Continuum.

Adkins, L. 2004. Passing on Feminism: From Consciousness to Reflexivity? *European Journal of Women's Studies* 11, 4: 427-44.

Agarwal, B. 1992. The Gender and Environment Debate: Lessons from India. *Feminist Studies* 181: 119-58.

Agrawal, A. 2005. *Environmentality: Technologies of Government and the Making of Subjects.* Durham, NC: Duke University Press.

Ahmed, S. 1998. *Differences That Matter: Feminist Theory and Postmodernism.* Cambridge: Cambridge University Press.

–. 2008. Imaginary Prohibitions: Some Preliminary Remarks on the Founding Gestures of the "New Materialism." *European Journal of Women's Studies* 15, 1: 23-39.

–. 2010. Orientations Matter. In Coole and Frost, *New Materialisms: Ontology, Agency, and Politics,* 234-57.

Alaimo, S. 2000. *Undomesticated Ground: Recasting Nature as Feminist Space.* Ithaca, NY: Cornell University Press.

Alaimo, S., and S.J. Hekman. 2008a. Introduction: Emerging Models of Materiality in Feminist Theory. In Alaimo and Hekman, *Material Feminisms,* 1-19.

–, eds. 2008b. *Material Feminisms.* Bloomington: Indiana University Press.

Alcoff, L. 1988. Cultural Feminism versus Post-Structuralism: The Identity Crisis in Feminist Theory. *Signs* 13, 3: 405-36.

Alexander, M.J., and C.T. Mohanty. 1997. *Feminist Genealogies, Colonial Legacies, Democratic Futures.* New York: Routledge.

Andermahr, S., T. Lovell, and C. Wolkowitz, eds. 2000. *A Glossary of Feminist Theory.* London: Arnold.

Andersen, M. 1993. *Women of the West Coast: Then and Now.* Sidney, BC: Sand Dollar Press.

Atwood, M. 1972. *Survival: A Thematic Guide to Canadian Literature.* Toronto: Anansi Press.

–. 1995. *Strange Things: The Malevolent North in Canadian Literature.* Oxford: Clarendon Press.

Bailey, A. 1994. Mothering, Diversity, and Peace Politics. *Hypatia* 92: 188-98.

Baker, S. 1993. *The Principles and Practice of Ecofeminism: A Review.* Rotterdam: Risbo.

Barad, K. 2003. Posthumanist Performativity: Towards an Understanding of How Matter Comes to Matter. *Signs* 28, 3: 801-31.

–. 2007. *Meeting the Universe Halfway: Quantum Physics and the Entanglement of Matter and Meaning.* Durham, NC: Duke University Press.

Barry, J. 2000. *Rethinking Green Politics.* London: Sage.

Bastian, M. 2011. The Contradictory Simultaneity of Being with Others: Exploring Concepts of Time and Community in the Work of Gloria Anzaldúa. *Feminist Review* 97, 1: 151-67.

Bell, S. 1993. Eco-Feminists Run "Peace Camp" at Clayoquot Sound. *Vancouver Sun,* 19 August, B1.

Bell, V. 1999. *Feminist Imagination: Genealogies in Feminist Theory.* London: Sage.

–. 2008. From Performativity to Ecology: On Judith Butler and Matters of Survival. *Subjectivity* 25, 1: 395-412.

Belmont, C. 2007. Ecofeminism and the Natural Disaster Heroine. *Women's Studies: An Inter-Disciplinary Journal* 36, 5: 349-72.

Berman, T. 1995. Standing for Our Lives: A Feminist Journey to Clayoquot Sound. Master's thesis, York University.

Berman, T., G.B. Ingram, M. Gibbons, R.B. Hatch, L. Maingon, and C. Hatch, eds. 1994. *Clayoquot and Dissent.* Vancouver: Ronsdale Press/Cacanadadada.

Beuge, D.J. 1994. Rethinking Again: A Defense of Ecofeminist Philosophy. In *Ecological Feminism,* ed. K.J. Warren, 42-63. London: Routledge.

Biehl, J. 1991. *Finding Our Way: Rethinking Ecofeminist Politics.* Montreal, Black Rose Books.

Bigwood, C. 1993. *Earth Muse: Feminism, Nature, and Art.* Philadelphia: Temple University Press.

Bly, R. 1992. *Iron John: A Book about Men.* Shaftesbury, UK: Element Books.

Bonnett, A. 1996. The New Primitives: Identity, Landscape and Cultural Appropriation in the Mythopoetic Men's Movement. *Antipode* 28, 3: 273-91.

Bossin, B. 1981. *Settling Clayoquot.* Victoria: Province of British Columbia, Ministry of Provincial Secretary and Government Services, Provincial Archives, Sound and Moving Image Division.

–. 1999. The Clayoquot Women. Bob Bossin's Old Folksinger's Home Page. http://www3.telus.net/oldfolk/women.htm.

Boucher, P.M. 1997. Ecology, Feminism, and Planning: Lessons from Women's Environmental Activism in Clayoquot Sound. PhD diss., University of British Columbia.

Bowker, G.C. 2005. *Memory Practices in the Sciences.* Cambridge, MA: MIT Press.

Braun, B. 2002. *The Intemperate Rainforest: Nature, Culture and Power on Canada's West Coast.* Minneapolis: University of Minnesota Press.

Breen-Needham, H., S.F. Duncan, D. Ferens, P. Reeve, and S. Yates, eds. 1994. *Witness to Wilderness: The Clayoquot Sound Anthology.* Vancouver: Arsenal Pulp Press.

Breton, M.J. 1998. *Women Pioneers for the Environment.* Boston: Northeastern University Press.

Brown, J. 1995. *All Possible Worlds: Utopian Experiments in British Columbia.* Vancouver: New Star Books.

Brown, P., and F.I.T. Ferguson. 1995. "Making a Big Stink": Women's Work, Women's Relationships and Toxic Waste Activism. *Gender and Society* 9, 2: 145-72.

Burman, E., and J. Stacey. 2010. The Child and Childhood in Feminist Theory. *Feminist Theory* 11, 3: 227-40.

Burton, A. 1992. "History" Is Now: Feminist Theory and the Production of Historical Feminisms. *Women's History Review* 1, 1: 25-39.

–. 1994. *Burdens of History: British Feminists, Indian Women, and Imperial Culture, 1865-1915.* Chapel Hill: University of North Carolina.

–. 2001. Thinking beyond the Boundaries: Empire, Feminism and the Domains of History. *Social History* 26, 1: 60-71.

Butler, J. 1990. *Gender Trouble: Feminism and the Subversion of Identity.* London: Routledge.

–. 1992. Contingent Foundations: Feminism and the Question of "Postmodernism." In *Feminists Theorize the Political,* ed. J. Butler and J. Scott, 3-21. London: Routledge.

–. 1993. *Bodies That Matter: On the Discursive Limits of "Sex."* London: Routledge.

–. 1994. Gender as Performance – Interview with Peter Osborne and Lynn Segal. *Radical Philosophy* 67: 32-39.

–. 2004. *Undoing Gender.* London: Routledge.

Butler, J., and J. Scott, eds. 1992. *Feminists Theorize the Political.* London: Routledge.

Byrne, B. 2003. Reciting the Self: Narrative Representations of the Self in Qualitative Interviews. *Feminist Theory* 4, 1: 29-49.

Caldecott, L., and S. Leland. 1983a. Introduction. In Caldecott and Leland, *Reclaim the Earth: Women Speak Out for Life on Earth,* 5-8.

–, eds. 1983b. *Reclaim the Earth: Women Speak Out for Life on Earth.* London: Women's Press.

Carlassare, E. 1994. Essentialism in Ecofeminist Discourse. In Merchant, *Ecology: Key Concepts in Critical Theory,* 220-34.

Carr, B. 1991. Between Continuity and Difference: An Interview with Daphne Marlatt. *West Coast Line* 25, 1: 99-107.

Carr, E. 1946. *Growing Pains: An Autobiography.* Toronto: Oxford University Press.

Castañeda, C. 2001. The Child as a Feminist Figuration: Toward a Politics of Privilege. *Feminist Theory* 2, 1: 29-53.

Castree, N. 2005. *Nature.* London: Routledge.

Chaloupka, W. 2003. There Must Be Some Way Out of Here: Strategy, Ethics, and Environmental Politics. In Magnusson and Shaw, *A Political Space: Reading the Global through Clayoquot Sound,* 67-90.

Chawla, L. 1994. *In the First Country of Places: Nature, Poetry, and Childhood Memory.* Albany: State University of New York Press.

–. 2002. *Growing Up in an Urbanising World.* London: Earthscan.

Chaudhuri, N. and M. Strobel, Eds. 1992. *Western Women and Imperialism: Complicity and Resistance.* Bloomington, Indiana University Press.

Cheney, J. 1994. Nature/Theory/Difference: Ecofeminism and the Reconstruction of Environmental Ethics. In *Ecological Feminism,* ed. K.J. Warren, 158-78. London: Routledge.

Chodorow, N., and S. Contratto. 1982. The Fantasy of the Perfect Mother. In Thorne and Yalom, *Rethinking the Family: Some Feminist Questions,* 54-75.

Clayton, D. 1996. Captain Cook and the Spaces of Contact at "Nootka Sound." In *Reading beyond Words: Contexts for Native History,* ed. J.S.H. Brown and E. Vibert, 95-123. Peterborough: Broadview.

–. 2000. *Islands of Truth: The Imperial Fashioning of Vancouver Island.* Vancouver: UBC Press.

Cobb, E. 1977. *The Ecology of Imagination in Childhood.* London: Routledge and Kegan Paul.

Cohen, M.P. 1996. Comment: Resistance to Wilderness. *Environmental History* 1, 1: 33-42.

Collard, A., and J. Contrucci. 1988. *Rape of the Wild: Man's Violence against Animals and the Earth.* London: Women's Press.

Collins, P.H. 1990. *Black Feminist Thought: Knowledge, Consciousness, and the Politics of Empowerment.* Boston: Unwin Hyman.

Coole, D.H., and S. Frost, eds. 2010. *New Materialisms: Ontology, Agency, and Politics.* Durham, NC: Duke University Press.

Cronon, W. 1992. A Place for Stories: Nature, History, and Narrative. *Journal of American History* 78, 4: 1347-76.

–. 1996. The Trouble with Wilderness: Or, Getting Back to the Wrong Nature. *Environmental History* 1, 1: 7-28.

Cuomo, C.J. 1998. *Feminism and Ecological Communities: An Ethic of Flourishing.* London: Routledge.

Cuthbertson, B., M.K. Heine, and D. Whitson. 1997. Producing Meaning through Movement: An Alternative View of Sense of Place. *The Trumpeter* 14, 2: 72-75.

Cvetkovich, A. 2003. *An Archive of Feelings: Trauma, Sexuality, and Lesbian Public Cultures.* Durham, NC: Duke University Press.

Daly, M. 1978. *Gyn/Ecology: Towards a Metaethics of Radical Feminism.* London: Women's Press.

Darier, É. 1996. Environmental Governmentality: The Case of Canada's Green Plan. *Environmental Politics* 5, 4: 585-606.

–, ed. 1999. *Discourses of the Environment.* Oxford: Blackwell.

Davion, V. 1994. Is Ecofeminism Feminist? In *Ecological Feminism,* ed. K.J. Warren, 8-28. London: Routledge.

Davis, N. 2009. New Materialism and Feminism's Anti-Biologism. *European Journal of Women's Studies* 16, 1: 67-80.

de Beauvoir, S. 1953. *The Second Sex.* London: Picador.

de Lauretis, T. 1989. The Essence of the Triangle or, Taking the Risk of Essentialism Seriously: Feminist Theory in Italy, the US and Britain. *differences: A Journal of Feminist Cultural Studies* 2: 3-37.

–. 1990. Upping the Anti (Sic) in Feminist Theory. In *Conflicts in Feminism,* ed. M. Hirsch and E.F. Keller, 255-70. London: Routledge.

–. 1993. Feminist Genealogies: A Personal Itinerary. *Women's Studies International Forum* 164: 393-403.

–. 1994. The Essence of the Triangle or, Taking the Risk of Essentialism Seriously: Feminist Theory in Italy, the U.S. and Britain. In *The Essential Difference,* ed. N. Schor and E. Weed, 1-39. Bloomington: Indiana University Press.

d'Eaubonne, F. 1974. *Le féminisme ou la mort.* Paris: Pierre Horay.

–. 1980. Feminism or Death. In Marks and De Courtivron, *New French Feminisms: An Anthology,* 64-67.

–. 1994. The Time for Ecofeminism. In Merchant, *Ecology: Key Concepts in Critical Theory,* 174-97.

Deleuze, G., and F. Guattari. 2004. *A Thousand Plateaus: Capitalism and Schizophrenia.* London: Continuum.

Delphy, C. 1984. *Close to Home: A Materialist Analysis of Women's Oppression.* London: Hutchinson.

Demeritt, D. 2001. Scientific Forest Conservation and the Statistical Picturing of Nature's Limits in the Progressive Era United States. *Environment and Planning D: Society and Space* 19: 431-59.

Diamond, I., and G.F. Orenstein. 1990a. Introduction. In Diamond and Orenstein, *Reweaving the World: The Emergence of Ecofeminism,* ix-xv.

–, eds. 1990b. *Reweaving the World: The Emergence of Ecofeminism.* San Francisco: Sierra Club Books.

Diamond, I., and L. Quinby, eds. 1988. *Feminism and Foucault: Reflections on Resistance.* Boston: Northeastern University.

DiChiro, G. 1992. Defining Environmental Justice: Women's Voices and Grassroots Politics. *Socialist Review* 22, 4: 93-130.

Dietz, M.G. 1985. Citizenship with a Feminist Face: The Problem of Maternal Thinking. *Political Theory* 13, 1: 19-37.

–. 1987. Context Is All: Feminism and Theories of Citizenship. *Daedalus* 116, 4: 1-24.

Di Prima, D. 2007. *Revolutionary Letters.* San Francisco: Last Gasp. (Orig. pub. 1971.)

Doherty, B., and M. d. Geus. 1996. Introduction. In *Democracy and Green Political Thought: Sustainability, Rights and Citizenship,* ed. B. Doherty and M. de Geus, 1-14. London: Routledge.

Drushka, K. 1995. *H.R.: A Biography of H.R. MacMillan.* Madeira Park, BC: Harbour.

Eagleton, M. 1996. *Working with Feminist Criticism.* Oxford: Blackwell.

Eaton, H. 1997. Liaison or Liability: Weaving Spirituality into Ecofeminist Politics. *Atlantis: A Women's Studies Journal* 21, 1: 109-22.

Eaton, H., and L.A. Lorentzen. 2003. Introduction. In *Ecofeminism and Globalization: Exploring Culture, Context, and Religion,* ed. H. Eaton and L.A. Lorentzen, 1-7. Lanham, MD: Rowman and Littlefield.

Echols, A. 1989. *Daring to Be Bad.* Minneapolis: University of Minnesota Press.

Edelman, L. 2004. *No Future: Queer Theory and the Death Drive.* Durham, NC: Duke University Press.

Eichhorn, K. 2010. D.I.Y. Collectors, Archiving Scholars, and Activist Librarians: Legitimizing Feminist Knowledge and Cultural Production since 1990. *Women's Studies: An Inter-Disciplinary Journal* 39, 6: 622-46.

Elshtain, J.B. 1981. *Public Man, Private Woman: Women in Social and Political Thought.* Princeton, NJ: Princeton University Press.

–. 1982. *The Family in Political Thought.* Amherst: University of Massachusetts Press.

Elshtain, J.B., and S. Tobias. 1990. *Women, Militarism, and War: Essays in History, Politics, and Social Theory.* Savage, MD: Rowman and Littlefield.

Emberley, J. 2007. *Defamiliarizing the Aboriginal: Cultural Practices and Decolonization in Canada.* Toronto: University of Toronto Press.

Enloe, C.H. 1983. *Does Khaki Become You? The Militarisation of Women's Lives.* London: Pluto.

Environment Canada. 1993. *A Primer on Environmental Citizenship.* Ottawa: Environment Canada. http://infohouse.p2ric.org/ref/03/02426.htm.

Eschle, C. 2004. "Skeleton Women": Feminism and the Antiglobalization Movement. *Signs* 30, 3: 1741-69.

Fairbarns, Z. 2003. Saying What We Want: Women's Liberation and the Seven Demands. In *The Feminist Seventies,* ed. H. Graham, A. Kaloski, A. Neilson, and E. Robertson, 93-103. York, UK: Raw Nerve Books.

Faludi, S. 1992. *Backlash: The Undeclared War against Women.* London: Vintage.

Felesky, L. 2001. Will Women Save the Earth? *Herizons.* http://www.herizons.ca/node/118.

Firestone, S. 1970. *The Dialectic of Sex: The Case for Feminist Revolution.* London: Women's Press.

Foucault, M. 1984. Nietzsche, Genealogy, History. In *The Foucault Reader.* Ed. by P. Rabinow, 76-100. Harmondsworth: Penguin.

Franklin, S. 2007. *Dolly Mixtures: The Remaking of Genealogy.* Durham, NC: Duke University Press.

Freeman, M., ed. 1995. *Always Rachel: The Letters of Rachel Carson and Dorothy Freeman, 1952-1964.* Boston: Beacon Press.

Freudenberg, N., and E. Zaltzberg. 1984. From Grassroots Activism to Political Power: Women Organizing against Environmental Hazards. In *Double Exposure: Women's Health Hazards on the Job and at Home,* ed. W. Chavkin, 246-47. New York, Monthly Review Press.

Fuss, D. 1989. *Essentially Speaking: Feminism, Nature, Difference.* London: Routledge.

Gaard, G. 1993a. Ecofeminism and Native American Cultures: Pushing the Limits of Cultural Imperialism? In Gaard, *Ecofeminism: Women, Animals, Nature,* 295-314.

–, ed. 1993b. *Ecofeminism: Women, Animals, Nature.* Philadelphia: Temple University Press.

–. 1998. *Ecological Politics: Ecofeminists and the Greens.* Philadelphia: Temple University Press.

–. 2002. Vegetarian Ecofeminism – A Review Essay. *Frontiers – A Journal of Women Studies* 23, 3: 117-46.

–. 2007. *The Nature of Home: Taking Root in a Place.* Tucson: University of Arizona Press.

–. 2011. Ecofeminism Revisited: Rejecting Essentialism and Re-Placing Species in a Materialist Feminist Environmentalism. *Feminist Formations* 23, 2: 26-53.

Gaard, G., and L. Gruen. 1993. Ecofeminism: Towards Global Justice and Planetary Health. *Society and Nature* 2, 1: 1-35.

Garb, Y. 1997. Lost in Translation: Towards a Feminist Account of Chipko. In *Transitions, Environments,Translations: Feminism in International Politics,* ed. J.W. Scott, C. Kaplan, and D. Keates, 273-84. London: Routledge.

Garland, A.W. 1988. *Women Activists: Challenging the Abuse of Power.* New York: The Feminist Press at the City University of New York.

Gatens, M. 1996. Sex, Contract and Genealogy. *Journal of Political Philosophy* 41: 29-44.

Gates, B.T. 1998. A Root of Ecofeminism: *Ecoféminisme.* In *Ecofeminist Literary Criticism: Theory, Interpretation, Pedagogy,* ed. G. Gaard and P. Murphy, 15-22. Urbana: University of Illinois Press.

George, E.M. 2003. *Living on the Edge: Nuu-chah-nulth History from an Ahousaht Chief's Perspective.* Winlaw, BC: Sono Nis.

Gibbs, L. 1997. What Is Your Wife Trying to Do – Shut Down the Chemical Industry? The Housewives of Love Canal. In *The Politics of Motherhood: Activist Voices from Left to Right,* ed. A. Jetter, A. Orleck, and D. Taylor, 28-43. Hanover, NH: University Press of New England.

Gibbs, L.M. 1998. *Love Canal: The Story Continues.* Gabriola Island, BC: New Society.

Gibson-Graham, J.K., and G. Roelvink. 2010. An Economic Ethics for the Anthropocene. *Antipode* 41, S1: 320-46.

Ginn, F. 2008. Extension, Subversion, Containment: Eco-Nationalism and (Post) Colonial Nature in Aotearoa New Zealand. *Transactions of the Institute of British Geographers* 33, 3: 335-53.

Griffin, S. 1978. *Woman and Nature: The Roaring inside Her.* New York: Harper and Row.

Grosz, E. 2004. *The Nick of Time: Politics, Evolution, and the Untimely.* Durham, NC: Duke University Press.

–. 2005. *Time Travels: Feminism, Nature, Power.* Durham, NC: Duke University Press.

–. 2010. The Untimeliness of Feminist Theory. *NORA – Nordic Journal of Feminist and Gender Research* 18, 1: 48-51.

Guppy, W. 1997. *Clayoquot Soundings: A History of Clayoquot Sound, 1800s to 1980s*. Tofino, BC: Grassroots.

Hagan, J. 2001. *Northern Passage: American Vietnam War Resisters in Canada*. Cambridge, MA: Harvard University Press.

Haig-Brown, A. 1996. *Hell No, We Won't Go: Vietnam Draft Resisters in Canada*. Vancouver: Raincoast Books.

Halberstam, J. 1998. *Female Masculinity*. Durham, NC: Duke University Press.

Hallstein, D.L.O.B. 2007. Matrophobic Sisters and Daughters: The Rhetorical Consequences of Matrophobia in Contemporary White Feminist Analyses of Maternity. *Women's Studies: An Inter-Disciplinary Journal* 36, 4: 269-96.

–. 2010. The Intriguing History and Silences of Of Woman Born: Rereading Adrienne Rich Rhetorically to Better Understand the Contemporary Context. *Feminist Formations* 22, 2: 18-41.

Hamilton, C. 1990. Women, Home and Community: The Struggle in an Urban Environment. In Diamond and Orenstein, *Reweaving the World: The Emergence of Ecofeminism*, 215-22.

Hampson, F.O. 1990. Pollution across Borders: Canada's International Environmental Agenda. In *Canada among Nations 1989: The Challenge of Change*, ed. M.A. Molot and F.O. Hampson, 175-92. Ottawa: Carleton University Press.

Haran, J. 2003. Re-Visioning Feminist Futures: Literature as Social Theory. PhD diss., University of Warwick.

–. 2010. Redefining Hope as Praxis. *Journal for Cultural Research* 14, 4: 393-408.

–. 2013. Screening NatureCultures: Eco/feminism and Green Film Production. Paper presented at "Naturecultures in Feminist Academia, Art and Activism," Utrecht University, Netherlands, 29 August.

–. 2014. Staying with the Trouble with Cyborgs and Goddesses. *The WisCon Chronicles Volume 7*, special issue edited by R. Holden. Seattle: Aqueduct Press, forthcoming.

Haran, J., and N. Moore. 2008. Revisioning Feminism: Imagining Feminist Futures. Paper presented at "Ending International Feminist Futures? Reviewing Sex, Gender and International Politics Conference," University of Aberdeen, 24-25 October.

Haraway, D. 1985. A Manifesto for Cyborgs: Science, Technology and Socialist Feminism in the 1980s. *Socialist Review* 80: 64-107.

–. 1988. Situated Knowledges: The Science Question in Feminism and the Privilege of Partial Perspective. *Feminist Studies* 14, 3: 575-99.

–. 1992. The Promises of Monsters: A Regenerative Politics for Inappropriate/d Others. In *Cultural Studies*, ed. L. Grossberg, C. Nelson, and P.A. Treichler, 295-337. London: Routledge.

–. 1997. *Modest_Witness@Second_Millennium.FemaleMan©_Meets_OncoMouse™: Feminism and Technoscience*. London: Routledge.

–. 2003. *The Companion Species Manifesto: Dogs, People and Significant Otherness*. Chicago: Prickly Paradigm Press.

–. 2004. *The Haraway Reader*. New York: Routledge.

–. 2008. *When Species Meet*. Minneapolis: University of Minnesota Press.

–. 2013. Sowing Worlds: A Seed Bag for Terraforming with Earth Others. In *Beyond the Cyborg: Adventures with Donna Haraway*, ed. M. Grebowicz and H. Merrick, 137-46. New York: Columbia University Press.

Hawkesworth, M. 2004. The Semiotics of Premature Burial: Feminism in a Post-feminist Age. *Signs* 29, 4: 961-85.

Hays, S.P. 1996. Comment: The Trouble with Bill Cronon's Wilderness. *Environmental History* 1, 1: 29-32.

Heath, S. 1978. Difference. *Screen* 19, 3: 51-112.

Hekman, S.J. 1996. *Feminist Interpretations of Michel Foucault*. University Park: Pennsylvania State University Press.

Heller, C. 1999. *Ecology of Everyday Life: Rethinking the Desire for Nature*. Montreal: Black Rose Books.

Helmreich, S. 2007. Induction, Deduction, Abduction, and the Logics of Race and Kinship. *American Ethnologist* 34, 2: 230-32.

Hemmings, C. 2005. Telling Feminist Stories. *Feminist Theory* 6, 2: 115-39.

–. 2011. *Why Stories Matter: The Political Grammar of Feminist Theory*. Durham, NC: Duke University Press.

Hessing, M., R. Raglon, and C. Sandilands. 2005. Environmental Politics: Issues at Home and Away. In *This Elusive Land: Women and the Canadian Environment*, ed. M. Hessing, R. Raglon, and C. Sandilands, 163-68. Vancouver: UBC Press.

Hird, M. 2009. Feminist Engagements with Matter. *Feminist Studies* 35, 2: 329-346.

Hird, M.J. 2004. Feminist Matters. *Feminist Theory* 5, 2: 223-32.

Hynes, H.P. 1985. Ellen Swallow, Lois Gibbs and Rachel Carson: Catalysts of the American Environmental Movement. *Women's Studies International Forum* 8, 4: 291-98.

–. 1989. *The Recurring Silent Spring*. New York: Pergamon Press.

Ingram, G.B. 1994. The Ecology of a Conflict. In Berman et al., *Clayoquot and Dissent*, 9-71.

Jackson, C. 1993a. Environmentalisms and Gender Interests in the Third World. *Development and Change* 24: 649-77.

–. 1993b. Women/Nature or Gender/History? A Critique of Eco/feminist Development. *Journal of Peasant Studies* 20, 3: 389-419.

–. 1995. Radical Environmental Myths: A Gender Perspective. *New Left Review* 210: 124-40.

Jaggar, A. 1983. *Feminist Politics and Human Nature*. Totowa, NJ: Rowman and Allanheld.

James, S. 1990. Is There a "Place" for Children in Geography? *Area* 22, 3: 278-83.

Jardine, A. 1987. Men in Feminism: Odo di Uomo or Compagnons de Route. In *Men in Feminism*, ed. A. Jardine and P. Smith, 54-61. London: Metheun.

Jones, O. 1997. Little Figures, Big Shadows: Country Childhood Stories. In *Contested Countryside Cultures: Otherness, Marginalisation, and Rurality*, ed. P.J. Cloke and J. Little, 152-72. London: Routledge.

–. 1999. Tomboy Tales: The Rural, Nature and the Gender of Childhood. *Gender, Place and Culture* 6, 2: 117-36.

Kaplan, C. 1994. The Politics of Location as Transnational Feminist Practice. In *Scattered Hegemonies: Postmodernity and Transnational Feminist Practices,* ed. I. Grewal and C. Kaplan, 137-52. Minneapolis: University of Minnesota Press.

Kaplan, T. 1997. *Crazy for Democracy: Women in Grassroots Movements.* New York: Routledge.

Kavka, M. 2001. Introduction. In *Feminist Consequences: Theory for the New Century,* ed. E. Bronfen and M. Kavka, ix-xxvi. New York: Columbia University Press.

–. 2002. Feminism, Ethics, and History, or What Is the "Post" in Postfeminism? *Tulsa Studies in Women's Literature* 21, 2: 29-44.

Kelly, P. 1989. Foreward: Linking Arms, Dear Sisters, Brings Hope! In *Healing the Wounds: The Promise of Ecofeminism,* ed. J. Plant, ix-xi. Philadelphia: New Society.

Kennedy, D. 1996. *The Garden Club and the Kumquat Campaign: A Novel.* Vancouver: Whitecap Books.

Kheel, M. 1991. Ecofeminism and Deep Ecology: Reflections on Identity and Difference. *The Trumpeter* 8, 2: 62-72.

King, K. 1994. *Theory in Its Feminist Travels: Conversations in US Women's Movements.* Bloomington: Indiana University Press.

–. 2010. Pastpresents: Playing Cat's Cradle with Donna Haraway. *Thinking with Donna Haraway.* http://partywriting.blogspot.com/.

Kirby, V. 2008. Natural Convers(at)ions: Or, What If Culture Was Really Nature All Along? In Alaimo and Hekman, *Material Feminisms,* 214-36.

Kogawa, J. 1981. *Obasan.* Toronto: Lester and Orpen Dennys.

–. 1992. *Itsuka.* Toronto: Viking.

Kossew, S. 2003. History and Place: An Interview with Daphne Marlatt. *Canadian Literature/Littérature canadienne* 178: 49-56.

Koven, S., and S. Michel. 1993. *Mothers of a New World: Maternalist Politics and the Origins of Welfare States.* London: Routledge.

Krauss, C. 1993. Women and Toxic Waste Protests: Race, Class and Gender as Resources of Resistance. *Qualitative Sociology* 163: 247-62.

Krawczyk, B. 1996. *Clayoquot: The Sound of My Heart.* Victoria: Orca.

Kurth-Schai, R. 1997. Ecofeminism and Children. In Warren, *Ecofeminism: Women, Culture, Nature,* 193-212.

Lahar, S. 1991. Ecofeminist Theory and Grassroots Politics. *Hypatia* 6, 1: 28-45.

Langer, V. 1994. It Happened Suddenly (Over a Long Period of Time): A Clayoquot History. In Breen-Needham et al., *Witness to Wilderness: The Clayoquot Sound Anthology,* 251-55.

Langer, V., and J. Bate. 1993. Women Out Front in Clayoquot Sound. In Women and Environments. Special issue, *Canadian Woman Studies/les cahiers de la femme* 13, 3: 81-83.

Lay, J. 1980. To Columbia on the Tynemouth: The Emigration of Single Women and Girls in 1862. In *In Her Own Right: Selected Essays in Women's History in B.C.,* ed. B. Latham and C. Kess, 19-41. Victoria: Camosun College.

Lean, G. 1991. Canadian Chainsaw Massacre. *Observer Magazine,* 1 December, 14-21.

Lear, L. 1997. *Rachel Carson: Witness for Nature.* London: Allen Lane.

Lipschutz, R.D. 2001. Environmental History, Political Economy and Change: Frameworks and Tools for Research and Analysis. *Global Environmental Politics* 1, 3: 72-91.

Lorde, A. 1984. The Master's Tools Will Never Dismantle the Master's House. In A. Lorde, *Sister Outsider: Essays and Speeches,* 110-14. Trumansburg, NY: Crossing Press.

Louv, R. 2009. *Last Child in the Woods: Saving Our Children from Nature-Deficit Disorder.* London: Atlantic.

Lowther, C. 1997. *A Cabin in Clayoquot.* Gabriola, BC: Pacific Edge.

Luke, T. 1999. Environmentality as Green Governmentality. In Darier, *Discourses of the Environment,* 121-51.

Lutrin, N. De Bertrand. 1928. *The Pioneer Women of Vancouver Island, 1843-1866.* Victoria: Women's Canadian Club.

MacGregor, S. 2006. *Beyond Mothering Earth: Ecological Citizenship and the Politics of Care.* Vancouver: UBC Press.

MacGregor, S., and B. Szerszynski. 2003. Environmental Citizenship and the Administration of Life. Paper presented at "Citizenship and the Environment Workshop," Newcastle University, UK, 4-6 September.

MacIsaac, R., and A. Champagne. 1994. *Clayoquot Mass Trials: Defending the Rainforest.* Gabriola Island, BC: New Society.

MacKay, D. 1982. *Empire of Wood: The MacMillan Bloedel Story.* Vancouver: Douglas and McIntyre.

Mackey, E. 2000. "Death by Landscape": Race, Nature, and Gender in Canadian Nationalist Mythology. *Canadian Woman Studies/les cahiers de la femme* 20, 2: 125-30.

Magnusson, W. 2003. The Puzzle of the Political. In Magnusson and Shaw, *A Political Space: Reading the Global through Clayoquot Sound,* 1-20.

Magnusson, W., and K. Shaw. 2003a. Conclusion: Clayoquot and the Politics Beyond. In Magnusson and Shaw, *A Political Space: Reading the Global through Clayoquot Sound,* 263-85.

–, eds. 2003b. *A Political Space: Reading the Global through Clayoquot Sound.* Minneapolis: University of Minnesota Press.

–. 2003c. Preface. In Magnusson and Shaw, *A Political Space: Reading the Global through Clayoquot Sound,* vii-ix.

Mani, L. 2009. *SacredSecular: Contemplative Cultural Critique.* New Delhi: Routledge.

Maracle, L. 1996. *I Am Woman: A Native Perspective on Sociology and Feminism.* Vancouver: Press Gang.

Marchak, P. 1995. *Logging the Globe.* Montreal and Kingston: McGill-Queen's University Press.

Marks, E., and I. De Courtivron, eds. 1980. *New French Feminisms: An Anthology.* Amherst: University of Massachusetts Press.

Marlatt, D. 1990. *Ana Historic: A Novel of Discovery.* London: Women's Press.

Marlatt, D., and R. Minden. 2001. *Steveston.* Vancouver: Ronsdale.

Martin, W. 1984. Another View of the "City upon a Hill": The Prophetic Vision of Adrienne Rich. In *Women Writers and the City: Essays in Feminist Literary Criticism,* ed. S.M. Squier, 249-64. Knoxville: University of Tennessee Press.

McClintock, A. 1995. *Imperial Leather: Race, Gender and Sexuality in the Colonial Contest.* London: Routledge.

McLaren, J. 1994. *Spirits Rising: The Story of the Clayoquot Peace Camp, 1993.* Gabriola Island, BC: Pacific Edge.

McLaren, J., and H. Brown, eds. 1993. *The Raging Grannies Songbook.* Gabriola Island, BC: New Society.

McMillan, L. 2004. Telling a Critical Story: Alice Walker's In Search of Our Mothers' Gardens. *Journal of Modern Literature* 28, 1: 107-23.

McNeil, M. 2010. Post-Millenial Feminist Theory: Encounters with Humanism, Materialism, Critique, Nature, Biology and Darwin. *Journal of Cultural Research* 14, 4: 427-37.

McRobbie, A. 2009. *The Aftermath of Feminism: Gender, Culture and Social Change.* London: Sage.

Meigs, M. 1991. *In the Company of Strangers.* Talonbooks: Vancouver.

Mellor, M. 1992. *Breaking the Boundaries: Towards a Feminist Green Socialism.* London: Virago.

Merchant, C. 1981. Earthcare: Women and the Environmental Movement. *Environment* 23, 5: 13, 6-40.

–. 1992. *Radical Ecology: The Search for a Livable World.* London: Routledge.

–, ed. 1994. *Ecology: Key Concepts in Critical Theory.* Atlantic Highlands, NJ: Humanities Press.

–. 1996. *Earthcare: Women and the Environment.* New York: Routledge.

Midgley, C. 1998. *Gender and Imperialism.* Manchester: Manchester University Press.

Minh-ha, Trinh T. 1989. *Women, Native, Other: Writing, Postcoloniality and Feminism.* Bloomington: Indiana University Press.

Miranda, L., and P. Joe. 1993. How the Squamish Remember George Vancouver. In *From Maps to Metaphors: The Pacific World of George Vancouver,* ed. R. Fisher and H. Johnston, 3-5. Vancouver: UBC Press.

Mohanty, C. 1992. Feminist Encounters: Locating the Politics of Experience. In *Destabilising Theory, Contemporary Feminist Thought,* ed. M. Barrett and A. Phillips, 74-92. Stanford: Stanford University Press.

Moore, N. 2002. The Ecofeminist Politics of Clayoquot Sound, Canada: Theorising Activist Narratives. DPhil diss., University of Sussex.

–. 2003. Ecological Citizens or Ecoterrorists? Learning through Environmental Activism in Clayoquot Sound. In *Adult Learning, Citizenship and Community Voices: Exploring Community-Based Practice,* ed. P. Coare and R. Johnston 92-107. Leicester, UK: National Institute of Adult Continuing Education.

–. 2004. Ecofeminism as Third Wave Feminism? Essentialism, Activism and the Academy. In *Third Wave Feminism: A Critical Exploration,* ed. S. Gillis, G. Howie, and B. Munford, 227-39. London: Palgrave Macmillan.

–. 2007. Imagining Feminist Futures: The Third Wave, Postfeminism and Eco/feminism. In *Third Wave Feminism: A Critical Exploration,* 2nd ed., ed. S. Gillis, G. Howie, and B. Munford, 125-41. London: Palgrave Macmillan.

–. 2008. The Rise and Rise of Ecofeminism as a Development Fable: A Response to Melissa Leach's "Earth Mothers and Other Ecofeminist Fables: How a Strategic Notion Rose and Fell." *Development and Change* 39, 3: 461-75.

–. 2011. Eco/Feminism and Rewriting the Ending of Feminism: From the Chipko Movement to Clayoquot Sound. *Feminist Theory* 12, 1: 3-21.

Munt, S.R. 2009. Queer Spiritual Spaces. In *Queer Spiritual Spaces: Sexuality and Sacred Spaces,* ed. K. Browne, S.R. Munt, and A.K.T. Yip, 1-33. London: Ashgate.

Nash, C. 2002. Genealogical Identities. *Environment and Planning D: Society and Space* 20, 1: 27-52.

Nelson, L. 1990. The Place of Women in Polluted Places. In Diamond and Orenstein, *Reweaving the World: The Emergence ofEcofeminism,* 173-200.

Norwood, V. 1987. The Nature of Knowing: Rachel Carson and the American Environment. *Signs* 12, 4: 740-60.

Norwood, V., and J. Monk. 1987. *The Desert Is No Lady: Southwestern Landscapes in Women's Writing and Art.* New Haven: Yale University Press.

O'Reilly, A. 2004. Introduction. In *From Motherhood to Mothering: The Legacy of Adrienne Rich's Of Woman Born,* ed. A. O'Reilly, 1-23. New York: State University of New York.

Ottenberg, S. 1990. Thirty Years of Fieldnotes: Changing Relationships to the Text. In *Fieldnotes: The Makings of Anthropology,* ed. R. Sanjek, 139-60. Ithaca: Cornell University Press.

Paglia, C. 1990. *Sexual Personae: Art and Decadence from Nefertiti to Emily Dickinson.* London: Penguin.

Parker, J. 2001. Towards an Ecofeminist Ethics: A Critical Realist and Social Movements Approach. PhD diss., University of Sussex.

Perry, A. 2001. *On the Edge of Empire: Gender, Race and the Making of British Columia, 1849-1871.* Toronto: University of Toronto Press.

Phelan, S. 1993. Intimate Distance: The Dislocation of Nature in Modernity. In *In the Nature of Things: Language, Politics and the Environment,* ed. J. Bennett and W. Chaloupka, 44-62. Minneapolis: University of Minnesota Press.

Plant, J. 1989a. The Circle Is Gathering. In Plant, *Healing the Wounds: The Promise of Ecofeminism,* 242-53.

–, ed. 1989b. *Healing the Wounds: The Promise of Ecofeminism.* Philadelphia: New Society.

–. 1989c. Toward a New World: An Introduction. In Plant, ed., *Healing the Wounds: The Promise of Ecofeminism,* 1-4.

Plumwood, V. 1993. *Feminism and the Mastery of Nature.* London: Routledge.

–. 2002. *Environmental Culture: The Ecological Crisis of Reason.* London: Routledge.

–. 2008. Shadow Places and the Politics of Dwelling. *Australian Humanities Review: Eco-Humanities Corner* 44: 139-50.

Pratt, M.L. 1992. *Imperial Eyes: Travel Writing and Transculturation.* London: Routledge.

Pritchard, A. 1992. The Shapes of History in British Columbia Writing. *BC Studies* 93: 48-69.

Puig de la Bellacasa, M. 2010. Ethical Doings in Naturecultures. *Ethics, Place and Environment* 13, 2: 151-69.

Reed, M.G. 2003. *Taking Stands: Gender and the Sustainability of Rural Communities.* Vancouver: UBC Press.

Rich, A. 1979. When We Dead Awaken: Writing as Re-Vision (1971). In A. Rich, *On Lies, Secrets and Silence: Selected Prose, 1966-1978,* 33-49. New York: W.W. Norton.

–. 1987. *Blood, Bread and Poetry: Selected Prose, 1979-1985.* London: Virago.

–. 1995. *Of Woman Born: Motherhood as Experience and Institution.* New York: W.W. Norton.

Roach, C. 1991. Loving Your Mother: On the Woman-Nature Relation. *Hypatia* 6, 1: 46-59.

–. 2003. *Mother/Nature: Popular Culture and Environmental Ethics.* Bloomington: Indiana University Press.

Rosenberg, H. 1987. The Kitchen and the Multinational Corporation: An Analysis of the Links between the Household and Global Corporations. *Journal of Business Ethics* 63: 179-94.

Roseneil, S. 2000. *Common Women, Uncommon Practices: The Queer Feminisms of Greenham.* London: Cassell.

Rowell, A. 1996. To Cut or Not to Clear-Cut: A Question of Trees, Truth and Treason. In A. Rowell, *Green Backlash: Global Subversion of the Environment Movement,* 182-208. London: Routledge.

Roy, C. 2004. *The Raging Grannies: Wild Hats, Cheeky Songs and Witty Actions for a Better World.* Montreal: Black Rose Books.

Ruddick, S. 1989. *Maternal Thinking: Towards a Politics of Peace.* London: Women's Press.

Rutherford, P. 1999. The Entry of Life into History. In Darier, *Discourses of the Environment,* 37-62.

Salleh, A. 1991. Review of Vandana Shiva's *Staying Alive: Women, Ecology and Development. Hypatia* 6, 1: 206-14.

–. 1997. *Ecofeminism as Politics: Nature, Marx, and the Postmodern.* New York: Zed Books.

Sandilands, C. 1997. Mother Nature, the Cyborg, and the Queer: Ecofeminism and (More) Questions of Identity. *National Women's Studies Association Journal* 9, 3: 18-40.

–. 1999. *The Good-Natured Feminist: Ecofeminism and the Quest for Democracy.* Minneapolis: University of Minnesota Press.

–. 2000. Domestic Politics: Multiculturalism, Wilderness and the Desire for Canada. *Space and Culture* 2, 4-5: 169-86.

–. 2002. Lesbian Separatist Communities and the Experience of Nature. *Organization and Environment* 152: 131-63.

–. 2003. Between the Local and the Global: Clayoquot Sound and Simulacral Politics. In Magnusson and Shaw, *A Political Space: Reading the Global through Clayoquot Sound,* 139-67.

Sandoval, C. 1991. U.S. Third World Feminism: The Theory and Method of Oppositional Consciousness. *Genders* 10: 1-24.

Sawicki, J. 1991. *Disciplining Foucault: Feminism, Power and the Body.* London: Routledge.

Schneider, J.W. 2005. *Donna Haraway: Live Theory.* London: Continuum.

Schor, N. 1994. Introduction. In *The Essential Difference*, ed. N. Schor and E. Weed, vii-xvii. Bloomington: Indiana University Press.

Scott, C. 1990. *The Company of Strangers.* Montreal: National Film Board of Canada.

Scott, J. 1992. Experience. In Butler and Scott, *Feminists Theorize the Political,* 22-40.

Seager, J. 1993. *Earth Follies: Feminism, Politics and the Environment.* London: Green Print.

–. 2003. Rachel Carson Died of Breast Cancer: The Coming of Age of Feminist Environmentalism. *Signs* 28, 3: 945-92.

Seed, J. 2007. Beyond Anthropocentrism. In *Thinking Like a Mountain: Towards a Council of All Beings,* ed. J. Seed, J. Macy, P. Fleming, and A. Naess, 35-40. Gabriola Island, BC: New Society. (Orig. pub. 1988.)

Segal, L. 1987. *Is the Future Female? Troubled Thoughts on Contemporary Feminism.* London: Virago Press.

Shelley, M.W. 1818. *Frankenstein, or the Modern Prometheus.* London: Vintage, 2007.

Shiva, V. 1989. *Staying Alive: Women, Environment and Development.* London: Zed Books.

Shugar, D.R. 1995. *Separatism and Women's Community.* Lincoln: University of Nebraska Press.

Sibley, D. 1991. Children's Geographies: Some Problems of Representation. *Area* 23: 269-70.

–. 1995. Families and Domestic Routines: Constructing the Boundaries of Childhood. In *Mapping the Subject: Geographies of Cultural Transformation,* ed. S. Pile and N. Thrift, 123-37. London: Routledge.

Skeggs, B., ed. 1995. *Feminist Cultural Theory: Process and Production.* Manchester: Manchester University Press.

–. 1997. *Formations of Class and Gender: Becoming Respectable.* London: Sage.

Smart, C., B. Neale, and A. Wade. 2001. *The Changing Experience of Childhood: Families and Divorce.* Cambridge: Polity.

Spelman, E.V. 1988. *Inessential Woman: Problems of Exclusion in Feminist Thought.* London: Women's Press.

Spivak, G.C. 1987. *In Other Worlds: Essays in Cultural Politics.* New York: Metheun.

–. 1989. In a Word, Interview with Ellen Rooney. In Another Look at Essentialism. Special issue, *Differences: Journal of Feminist Cultural Studies* 1, 2: 124-56.

–. 1994. In a Word. Interview with Ellen Rooney. In *The Essential Difference,* ed. N. Schor and E. Weed, 151-82. Bloomington: Indiana University Press.

Springer, K. 2005. *Living for the Revolution: Black Feminist Organizations, 1968-1980.* Durham, NC: Duke University Press.

Starhawk. 1987. *Truth or Dare: Encounters with Power, Authority, and Mystery.* San Francisco: HarperSanFrancisco.

–. 1993. *The Fifth Sacred Thing.* New York: Bantam Books.

–. 1999. *The Spiral Dance: A Rebirth of the Ancient Religion of the Great Goddess.* San Francisco: Harper SanFrancisco.

–. 2002. *Webs of Power: Notes from the Global Uprising.* Gabriola Island, BC: New Society.

Stengers, I. 2008. Experimenting with Refrains: Subjectivity and the Challenge of Escaping Modern Dualism. *Subjectivity* 22, 1: 38-59.

Stephenson, N., and D. Papadopoulos. 2006. *Analysing Everyday Experience: Social Research and Political Change.* Basingstoke: Palgrave Macmillan.

Stone, A. 2004. Essentialism and Anti-Essentialism in Feminist Philosophy. *Journal of Moral Philosophy* 1, 2: 135-53.

–. 2005. Towards a Genealogical Feminism: A Reading of Judith Butler's Political Thought. *Contemporary Political Theory* 4: 4-24.

Sturgeon, N. 1997. *Ecofeminist Natures: Race, Gender, Feminist Theory and Political Action.* London: Routledge.

Sullivan, S. 2004. "We Are Heartbroken and Furious!" (#2) Violence and the (Anti-)Globalisation Movement(s). Centre for the Study of Globalisation and Regionalisation Working Paper no. 133/04. http://www2.warwick.ac.uk/fac/soc/csgr/research/abstracts/13304/.

Tamboukou, M. 2003. Writing Feminist Genealogies. *Journal of Gender Studies* 12, 1: 5-19.

Tennant, P. 1990. *Aboriginal People and Politics: The Indian Land Question in BC, 1849-1989.* Vancouver: UBC Press.

Thomashow, M. 1995. *Ecological Identity: Becoming a Reflective Environmentalist.* Cambridge, MA: MIT Press.

Thorne, B. 1982. Feminist Rethinking of the Family: An Overview. In Thorne and Yalom, *Rethinking the Family: Some Feminist Questions,* 1-24.

–. 1987. Re-Visioning Women and Social Change: Where Are the Children? *Gender and Society* 1, 1: 85-109.

Thorne, B., and M. Yalom, eds. 1982. *Rethinking the Family: Some Feminist Questions.* London: Longman.

Thorpe, J. 2012. *Temagami's Tangled Wild: Race, Gender, and the Making of Canadian Nature.* Vancouver: UBC Press.

Timpson, A.M., ed. 2009. *First Nations, First Thoughts: The Impact of Indigenous Thought in Canada.* Vancouver: UBC Press.

Toner, G., and T. Conway. 1996. Environmental Policy. In *Border Crossings: The Internationalization of Canadian Public Policy,* ed. G.B. Doern, L.A. Pal, and B.W. Tomlin, 108-44. Toronto: Oxford University Press.

Trebilcot, J., ed. 1984. *Mothering: Essays in Feminist Theory.* Totowa: Rowman and Littlefield.

Tsing, A.L. 1997. Transitions as Translations. In *Transitions, Environments, Translations: Meanings of Feminism in Contemporary Ethics,* ed. J.W. Scott, A. Kaplan, and D. Keates, 253-72. London: Routledge.

–. 2005. *Friction: An Ethnography of Global Connection.* Princeton: Princeton University Press.

Twine, R. 2010. Intersectional Disgust? Animals and (Eco)Feminism. *Feminism and Psychology* 20, 3: 397-406.

Unger, N. 2004. Women, Sexuality, and Environmental Justice in American History. In *New Perspectives on Environmental Justice: Gender, Sexuality, and Activism*, ed. R. Stein, 45-60. New Brunswick, NJ: Rutgers University Press.

Van der Tuin, I. 2008. Deflationary Logic: Response to Sara Ahmed's "Imaginary Prohibitions: Some Preliminary Remarks on the Founding Gestures of the 'New Materialism.'" *European Journal of Women's Studies* 15, 4: 411-16.

–. 2009. Jumping Generations: On Second- and Third-Wave Feminist Epistemology. *Australian Feminist Studies* 24, 59: 17-31.

Vance, L. 1993. Ecofeminism and the Politics of Reality. In Gaard, *Ecofeminism: Women, Animals, Nature*, 118-45.

Vidal, J. 1993. At Loggerheads. *Guardian Weekend* (London), 21 August, 1-30.

Visweswaran, K. 1994. *Fictions of Feminist Ethnography.* Minneapolis: University of Minnesota Press.

Walby, S. 2011. *The Future of Feminism.* Cambridge: Polity.

Walker, A. 1983. In Search of Our Mothers' Gardens. In A. Walker, *In Search of Our Mothers' Gardens*, 231-43. New York: Harcourt Brace Jovanovich.

Ward, C. 1978. *The Child in the City.* Oxford: Architectural Press.

Ware, V. 1992. *Beyond the Pale: White Women, Racism and History.* London: Verso.

Warren, K.J. 1987. Feminism and Ecology: Making Connections. *Environmental Ethics* 9: 3-20.

–. 1990. The Power and Promise of Ecological Feminism. *Environmental Ethics* 12, 2: 125-46.

–, ed. 1997a. *Ecofeminism: Women, Culture, Nature.* Bloomington: Indiana University Press.

–. 1997b. Taking Empirical Data Seriously: An Ecofeminist Philosophical Perspective. In Warren, *Ecofeminism: Women, Culture, Nature*, 3-20.

–. 1998. Towards an Ecofeminist Ethic. *Studies in the Humanities* 15: 140-56.

Watts, R. 1992. B.C.'s Handling of Logging Issue Has Latin Journalists Snickering. *Victoria Times-Colonist*, 15 December.

Weisberg, Z. 2009. The Broken Promises of Monsters: Haraway, Animals and the Humanist Legacy. *Journal for Critical Animal Studies* 7, 2: 22-62.

Whatmore, S. 2006. Materialist Returns: Practising Cultural Geography in and for a More-Than-Human World. *Cultural Geographies* 13, 4: 600-9.

Wheeler, S. 1994. How Far It Runs. In Breen-Needham et al., *Witness to Wilderness*, 28.

Wiegman, R. 2000. Feminism's Apocalyptic Futures. *New Literary History* 31: 805-25.

–. 2008. The Intimacy of Critique. Keynote presentation at "Ending International Feminist Futures? Reviewing Sex, Gender and International Politics Conference," University of Aberdeen, 24-25 October.

Williams, J.J. 2009. Science Stories: An Interview with Donna J. Haraway. *Minnesota Review* 2010, 73-74: 133-63.

Williams, L.R. 1992. Happy Families? Feminist Reproduction and Matrilineal Thought. In *New Feminist Discourses: Critical Essays on Theories and Texts*, ed. I. Armstrong, 48-64. London: Routledge.

Williams, T.T. 1991. *Refuge: An Unnatural History of Family and Place.* New York: Vintage Books.

Wilson, E.A. 2004. Gut Feminism. *differences: A Journal of Feminist Cultural Studies* 15, 3: 66-94.

–. 2010. *Affect and Artificial Intelligence.* Seattle: University of Washington Press.

Wine, S. 1997. *Fury for the Sound: The Women at Clayoquot.* Vancouver: TellTale Productions.

Wolf, N. 1991. *The Beauty Myth.* London: Vintage.

Wollstonecraft, M. 1792. *A Vindication of the Rights of Woman.* Harmondsworth: Penguin, 1978.

Woods, M., J. Anderson, S. Guilbert, and S. Watkin. 2013. Rhizomatic Radicalism and Arborescent Advocacy: A Deleuzo-Guattarian Reading of Rural Protest. *Environment and Planning D: Society and Space* 31, 3: 434-50.

Woolf, V. 2002. *A Room of One's Own.* Harmondsworth: Penguin. (Orig. pub. 1929.)

Index

Printed and bound in Canada by Friesens

Set in Segoe and Warnock by Artegraphica Design Co. Ltd.

Copy editor: Deborah Kerr